This book is dedicated to those pilots who have worked to improve the level of aviation safety. From the Wright Brothers to every flight instructor who trains students to fly safely and competently, the desire to gain experience and pass that along to others is a special heritage. Without this professional approach to aviation, we would not be where we are today.

I would also like to dedicate this book to Dottie, Ryan, and Emily. The time it takes to complete an undertaking of this size is significant, and it would not have been possible without their support and understanding.

A special thanks to Lee Johnson, who has proofed each of the books I have written and shows fortitude in continuing to fly with me.

Thanks also to the management and staff of Wisconsin Aviation, who strive to turn out the best pilots possible.

Flight
Maneuvers

Flight Maneuvers

Michael C. Love

McGraw-Hill

New York   San Francisco   Washington, D.C.   Auckland   Bogotá
Caracas   Lisbon   London   Madrid   Mexico City   Milan
Montreal   New Delhi   San Juan   Singapore
Sydney   Tokyo   Toronto

Library of Congress Cataloging-in-Publication Data

Love, Michael C.   (Michael Charles)
   Flight maneuvers / Michael C. Love.
      p.   cm.
   Includes index.
   ISBN 0-07-038866-0 (hardcover : alk. paper). — ISBN 0-07-038865-2
  (pbk. : alk. paper)
    1. Airplanes—Piloting.   I. Title.
  TL710.L68   1998
  629.132'52—dc21
                                         98-37266
                                          CIP

McGraw-Hill

A Division of The McGraw·Hill Companies

1 2 3 4 5 6 7 8 9 0   DOC/DOC   9 0 3 2 1 0 9 8

ISBN 0-07-038865-2 (PBK)
ISBN 0-07-038866-0 (HC)

The sponsoring editor for this book was Shelley Ingram Carr, the editing supervisor was Sally Glover, and the production supervisor was Sherri Souffrance. It was set in Times per the PFS design by Michele Pridmore of McGraw-Hill's Professional Group Composition Unit, in Hightstown, NJ.

Printed and bound by R. R. Donnelley & Sons Company.

McGraw-Hill books are available at special quantity discounts to use as premiums and sales promotions, or for use in corporate training programs. For more information, please write to the Director of Special Sales, McGraw-Hill, 11 West 19th Street, New York, NY 10011. Or contact your local bookstore.

 This book is printed on recycled, acid-free paper containing a minimum of 50 percent recycled, de-inked fiber.

Contents

CONTENTS

Introduction

During my flying career, which has spanned more than twenty years, I have always been interested in learning to do more than just drive an airplane from point A to point B. While for the majority of us this is exactly what we use an airplane for, there is a great deal more to becoming a competent, proficient pilot. While a portion of that interest has always been there for me, I also must give credit to my private and commercial flight instructors. They had the desire to teach me the subtleties of flying an airplane, from how the controls feel, to why it is important to keep the ball centered, to stalls, spins, and all manner of information in between.

Flying offers the opportunity to constantly hone our skills and improve not only what we know about flying, but also how well we fly. This book is written out of a desire to carry on the work my flight instructors began with me, not only to know how to fly, but how to fly safely and with a level of professionalism. A number of topics are discussed in this book, from very basic fight maneuvers to the more advanced. As the title implies, the content centers on topics that include slow flight, stalls, and ground reference maneuvers. We also review takeoffs, landings, and emergency flight maneuvers. In order to reach the highest levels of safety and professionalism, we must know more than how to "drive" the airplane. We must know how it will react under varying conditions; we must be able to anticipate the plane as we fly it. This skill level comes with knowledge and practice but is quite obtainable for pilots who set that as a goal.

As we review each topic, you will learn not only the mechanics of how to fly a maneuver, but also you will begin to see why an airplane behaves in various ways during flight maneuvers. My main reason for learning to fly all those years ago was to become involved in aerobatics. While it took more time to get to do that type of flying than I wanted, I took the opportunity to make every flight a learning experience, to absorb the feel of how the airplane reacts in a given circumstance. This basic foundation has carried over to aerobatic flying, where "feeling" the airplane is one of the most important things a pilot must do.

INTRODUCTION

The book is not intended to be a self-taught course in flying, but provides valuable information and insight into how to fly many of the maneuvers we learn as pilots. Combined with a competent flight instructor, you can use the book to improve your level of knowledge, flying ability, and safety. Whenever you fly, safety should be the most important criteria you judge your performance by, and throughout the book safety is an important topic. All too frequently we read of private and commercial pilots making an error in judgment that results in an unfortunate end to a flight. The number of hours the pilot has or the type of plane the pilot is flying are immaterial; poor judgment can cause any pilot to get into an emergency situation. To help avoid these situations, we also discuss the poor judgment chain. In almost every accident, the pilots have several opportunities to avoid the situations, but they ignore the warning signs and blindly fly on until no escape options are available. If the warning signs are recognized, pilots can in most cases avert a serious situation, but we must learn to put safety first in order to do that.

The importance of control coordination is discussed at length. During any slow flight maneuver, including stalls, takeoffs, and landings, you must maintain coordinated flight. This aids you in avoiding misuse of the controls and potential stall/spin accidents. Most aircraft accidents take place during the takeoff/landing phase of a flight. In the majority of cases these accidents are due to pilot error. National Transportation and Safety Boards (NTSB) reports document that most low-altitude stall/spin accidents are due to misuse of the controls and airplane. By learning how to maintain coordinated flight by "feel," you can fly the airplane more safely, especially in an emergency situation. As we discuss various aspects of flight, stall/spin avoidance will also be reviewed in depth. The best way to keep from getting into an accidental stall or spin is to be able to recognize that you are approaching one and take corrective action before onset.

Chapter 1 is a discussion of the basics of aerodynamics. The forces acting on an airplane during flight are discussed in detail. In addition, a thorough review of lift is also covered. Many pilots do not understand what effect these forces and lift have on how an airplane flies, so we begin by building the foundation with this knowledge. Airfoils and lift enhancement devices are also given detailed review. As you will see, the flight characteristics of an airplane can change dramatically when flaps and other lift enhancement devices are actuated. By knowing the effect on the airplane, you can anticipate how to fly the airplane when they are used. Since many of these devices are used to allow an airplane to fly slower, it becomes even more important to understand the correct use of controls during slow flight scenarios. Finally, the affect of flight controls—including elevators, rudder, and ailerons—are also covered in the chapter.

Basic flight maneuvers are covered in Chapter 2. This discussion includes straight-and-level flight, climbs, descents, turns, and steep turns. Proper control inputs, power settings, and how angle of attack can affect each of these maneuvers are also reviewed. Additionally, potential problem areas that can affect pilots are discussed. This includes misuse of the controls and the effect this can have, and mistakes that pilots commonly make when flying these maneuvers. We also review airspeed and altitude control and their relationship with each other. Many pilots have misconceptions about how the flight controls, including the throttle, affect airspeed and altitude. To fly the airplane well, especially at slow airspeeds, you must know how to use the controls correctly.

Chapter 3 provides in-depth coverage of slow flight and stalls. The basics that are covered in Chapter 2 are expanded on in the third chapter. The discussion begins with the principles of slow flight, control effectiveness as airspeed is reduced, left-turning tendencies at slow airspeeds, and how to correctly enter and recover from slow flight. We then move on to stalls, an area that many pilots today are unfortunately not as familiar with as they should be. We begin with a definition of stalls and how angle of attack is related to stall speed. The critical angle of attack is discussed to help readers understand why an airplane can stall even with the plane's nose does not seem very high in relation to the horizon. You will also be able to see how a plane flying along at cruise airspeeds can stall. Normal versus accelerated stalls will be compared as part of the review. Stall entry and recovery and stall avoidance techniques are covered as part of the chapter as well. Finally, the difference between approach and departure stalls is covered. Even when the engine is developing full power, if the plane is flown incorrectly, you can stall it. Your center of gravity can also set you up for stalls, leading to another area we review.

Chapter 4 focuses on spins, a potential outcome of stalls if the plane is not flown correctly during a stall situation. We define what a spin is and what causes the plane to spin. Under the right circumstances, doing spins is enjoyable. But spins at low altitudes, such as during takeoff or landing, can turn into a very dangerous scenario. The phases of a spin—spin entry, the incipient spin, the developed spin, and spin recovery—are given in-depth coverage. Developing a better understanding of the phases a spin progresses through can help you view spins in a different light. Today the majority of pilots never receive spin training. This results in misunderstanding and fear regarding spins. While unplanned spins can be dangerous, practicing spins in an approved aircraft, with a flight instructor who is knowledgeable in them, can be an enjoyable experience. Not only does this practice make you a more knowledgeable pilot, but also it can help you recognize the early warning signs of an imminent stall/spin and help you avoid them. Chapter 4 also covers a number of improper spin recovery procedures. The chapter ends with a review of several exotic spins, including inverted spins, flat spins, accelerated spin, and crossover spins. In the right aircraft, even these spins can be learned and practiced. But even if you never fly them, knowing what these spins are, and how to avoid them, can help you as you fly. Incorrect control inputs during a spin can cause the situation to become worse. Use of correct control inputs while in a spin will help you recover from the spin more quickly, and with less altitude loss. As you will see, even an hour of dual in spin training can help you better understand what a plane is doing while it spins, how a spin feels to you as you fly the airplane, and how to recover quickly from the spin. A large percentage of pilots I have given spin training to have never done a spin before. But even worse, some of the students I have trained have received only minimal spin training from an instructor that, I'm certain unintentionally, created an even greater fear of spins by projecting how serious and dangerous spin training is. While safety is always paramount in flying, an instructor that is nervous about a particular flight maneuver only increases the discomfort of the student.

INTRODUCTION

Chapter 5 begins a new area of discussion—ground reference maneuvers. As students, we all learn how to perform the basic ground reference maneuvers such as S-turns, turns about a point, and rectangular course. But once the rating has been tucked into our wallet, many of us never perform these maneuvers again. As a result we forget some of the lessons the maneuvers were designed to teach us. In this chapter we discuss the need for understanding ground reference maneuvers and how the wind affects our patterns during takeoff and landing. Too often pilots overshoot the turn from base to final because the wind is pushing them faster than they anticipated. The pilot often banks the plane more steeply or tries to rudder the nose around,both situations being equally dangerous. By learning how to anticipate what the wind's effect will be on the track you are flying over the ground, you can learn to fly more safely in winding conditions during takeoff and landing. Advanced ground reference maneuvers, eights around, and eights on pylon are reviewed in the last half of the chapter. Normally part of the commercial flight training curriculum, these maneuvers can help you learn how to account for the effects of wind in more complex situations, giving you an even better understanding of how to fly an airplane in the pattern on windy days.

Takeoffs and landings are covered in Chapter 6. The discussion begins with the various V speeds you should be familiar with. Knowing this information can help you better anticipate and control how the plane you fly will react, especially during takeoff and landing. Stall V speeds, maneuvering speed, and rate and angle of climb speeds are covered. By utilizing the V speeds correctly, you can improve the performance of the airplane you fly. The traffic pattern, and various entry and exit into it are next in the chapter. By following standard entry and exits to and from the pattern, you can improve safety and the flow of traffic in the pattern. We then move on to landing techniques. A landing is made up of several phases, and knowing what they are, and how to fly them, can improve your landings. Picking the touchdown spot and being able to consistently hit it is the sign of a pilot who understands how to land an airplane. Correctly judging your height, and timing your flares, can also make your landings feather light. Each of these areas, and others, are discussed in the chapter. Normal, short-field, soft-field and crosswind takeoffs and landings are also covered. Knowing how to safely execute each of these will improve your ability to fly the airplane within its capabilities. Common errors pilots make and abort procedures are also covered. The chapter discusses emergency landing procedures, helping you understand how to fly an airplane during a forced landing. While no one ever wants to experience one, knowing how to react when the engine stops can save the day. Remembering to fly the airplane, finding a suitable landing spot, and hitting it may all seem like common-sense tasks you would perform in an emergency-landing situation. But too often pilots forget what they should do, and they end up making mistakes that damage the plane, cause injury to them or their passengers, or worse.

Commercial maneuvers such as chandelles, lazy 8s, and steep spirals are covered in Chapter 7. The primary reason for learning these maneuvers is to learn to fly the airplane consistently well in changing flight conditions. Many of the maneuvers result in changing airspeeds and attitudes. Through proper use of the controls, you will begin to "feel" what the airplane is telling you. You will also learn how to compensate with the controls as the airspeed decreases or increases. Many of the aspects of slow flight you learned in Chapter 3 will apply directly to the maneuvers in this chapter. Once again, learning how to fly an airplane well in these varying flight conditions will improve your level of competence and safety.

Chapter 8 continues to discuss emergency procedures such as being lost, encountering unexpected bad weather, and other areas. While no book can outline every emergency, or how to react to it, you can build a basic understanding of what gets pilots into bad situations and how to avoid them. Quite frequently, emergencies could have been avoided if the pilot had recognized a potentially bad situation earlier, or reacted in a different manner to it. We often deny that a problem exists, which only makes the situation worse as the problem compounds itself. By recognizing early warning signs, you can in most cases avoid serious problems.

In *Spin Management and Recovery*, I devoted a chapter to National Transportation Review Board (NTSB) accident reports. As I reviewed the reports while preparing the chapter, the types of spin accidents pilots can get into amazed me. What seemed like common-sense actions to avoid an accident were all too often ignored. Stall/spins and fuel starvation are some of the most common accidents, yet you can avoid them by planning your flights and knowing how to fly your airplane. We also discuss this, as well as the pilot judgment chain, in Chapter 8. All too often, "get-home-itis" in a pilot results in poor decision making, in some cases the results being fatal. By breaking the chain of bad decisions early enough, pilots can recognize they are setting themselves up for an accident, and take the appropriate action. If the mistakes are realized too late, then an alternative may no longer be available. As you will see, our thinking is often obscured by factors such as fatigue, a desire to finish the flight, or wanting to get to an airport before weather hits. Learning to recognize when not to fly is one of the most important safety steps you can take.

Basic instrument flying is covered in Chapter 9. While whole texts can be written on the subject, we are going to examine the basic concepts of instrument flying. Every pilot should have learned how to fly the plane on instruments as part of their private pilot flight training. Too often, though, once pilots pass the exam and get a private license, they never get any refresher training. The purpose behind this chapter is to help you review some of the basics you should have learned but may not have thought about in a long time. Instrument function, the basic instrument scan, and how to maneuver on instruments are some of the topics we discuss. Finally, basic instrument navigation is covered. While you should never enter instrument flight rule (IFR) weather without a current instrument rating, knowing how to handle the plane on instruments can help you get out of clouds and back to clear skies.

The conclusion summarizes the major points made in the book. Aircraft control and how to anticipate how an airplane will fly in various flight regimes is reviewed. Your primary responsibility as a pilot, from preflight to touchdown, should be safety. By knowing your limitations and the limitations of the airplane you are flying, you can reduce the opportunity for problems to arise. Follow-on flight training is also discussed.

INTRODUCTION

Many pilots get into a routine of how and where they fly. When faced with a new situation, they may lack the experience to adequately handle difficult situations. By occasionally taking refresher training with a qualified flight instructor, pilots can hone their skills. We also discuss more advanced training, such as aerobatics. While not every pilot wants to know how to loop and roll an airplane, knowing what the world looks like when a plane is upside down can help keep you from becoming disoriented in an unusual attitude. Proper training in basic aerobatics can not only be fun, but also improve your abilities as a pilot. I hope you find that the information in this book is useful and that you take the time to follow up by practicing with a qualified flight instructor.

1
Basic Aerodynamics

T HE MAJORITY OF THIS BOOK DEALS WITH LEARNING HOW TO FLY A variety of maneuvers, from steep turns to slow flight to takeoffs and landings. The goal of the book is that when you are done reading, you will know more than just what control inputs to use during a maneuver. Instead, you should come away with a "feel" for the airplane, what it is telling you as you fly it. The sensitivity of the controls, the sound of air flowing over the plane, the noise the engine makes, all give you feedback about how the plane is flying. To help us get to that level of connection with the airplane, we are going to begin by discussing the basics of how an airplane flies—aerodynamics. Why we need to use rudder during turns, how an airplane reacts to the use of ailerons, and a number of other topics related to basic aerodynamics will be covered in this chapter. I have read other books that get into some pretty heavy equations in trying to explain aerodynamics, but we will not do that in this book. As opposed to lift over drag equations or detailed equations covering lift coefficients, I prefer to use examples to explain a particular idea. This is not a slam against the mathematics behind the aerodynamics, because this is very important to the engineers who design aircraft. My preference for teaching aerodynamics has centered more on examples; over the years this method has seemed to be well received by students I fly with.

As you read the chapter, keep in mind that the concepts presented are generalities; they may or may not apply to a particular make or model of aircraft. The best way you can approach this book, and flying in general, is to take the knowledge you have and adapt it to a given situation. There is no one rule that always applies; as the pilot you must be the ultimate judge of what your actions should be in a given situation. As you read, try to apply the information we cover to the flying you have already done. Does the plane you fly behave the same way as the examples in the book? If not, how does it differ, and why? As stated in the introduction, this is not a self-taught course in flying, so if you have questions, contact a qualified flight instructor to help you work through them. And keep in mind that the best classroom is the cockpit; being hands-on is the most efficient way to learn anything.

The starting point for our discussion of aerodynamics is centered on how an airplane flies, or generates lift. In this section we will cover the four forces acting on an airplane, define Bernoulli's principle, and cover the different types of lift. Following that we will look at several different airfoils and lift enhancement devices. The flight controls of an airplane and how an airplane actually turns will finish out the chapter.

FORCES ACTING ON AN AIRPLANE

As an airplane flies there are four forces that constantly act on it: lift, gravity, thrust, and drag (Figure 1-1). As you look at the figure, notice that lift and gravity oppose each other, as do thrust and drag. In order for a plane to get off the ground, the lift it is generating must be greater than the weight of the plane, or the pull of gravity on it. Once in the air, gravity is still trying to pull the plane back to the ground, so lift must be maintained for it to stay at a given altitude. When lift is reduced, such as in a descent, the plane loses altitude as a result of the force of gravity. In order to climb, the force of lift must be greater than the force of gravity. When a plane is at a constant altitude, lift and gravity equal each other.

Thrust and drag are also opposing forces acting on an airplane when it flies. In order to move forward, a plane must generate thrust. In most general-aviation aircraft, the

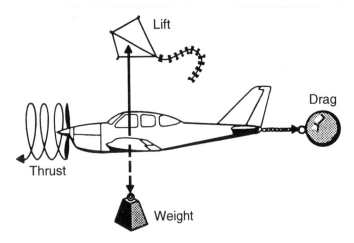

Fig. 1-1 *Four forces acting on an airplane.*

thrust is created by a propeller. (If you're really lucky, it can be generated by a jet engine.) The propeller is basically an airfoil turned by the engine; as it turns, it generates the thrust that moves an airplane forward through the air. You have probably noticed that some airplanes have propellers in the rear, known as pusher propellers, while the majority of them are located at the nose of the plane. These are known as tractor propellers because they pull the plane through the air. In both cases, the function of the propeller is the same—to generate the thrust needed to move the airplane forward.

Drag

As the plane moves forward, the motion through the air results in friction. Even though you cannot see air, it acts on an airplane in the same manner that water acts on your hand as you move it through the water. The friction, or *drag*, is the resistance of the air to a body moving through it, in this case a plane. The faster an object moves through the air, the greater the drag that is created by the friction. In fact, there are several types of drag created when an airplane flies. The two basic types of drag are parasite drag and induced drag. *Parasite drag* is the combined effects of form drag and skin friction. *Form drag* is the result of the disruption of a streamlined airflow and can be reduced by making the airplane as aerodynamically streamlined as possible during its design. Skin friction drag is caused by the resistance of airflow across the surface of an airplane and can be reduced by making the skin as smooth as possible. As you look over airplanes sitting on the ramp, notice that many rivets and screws on metal airplanes are flush mounted. This helps reduce the drag of air on the plane. If you get a chance to look at some of the higher-performance composite airplanes, you will notice there are very few rivets, and the skin of the plane is almost glass smooth. There is good reason for this; parasite drag increases roughly as the square of the increase in airspeed. For example, if you double your airspeed, the parasite drag of the plane will increase to four times the amount that was present at the original airspeed. As you can see, it is in the best interest of aircraft designers to keep them as aerodynamic as possible to reduce parasite drag.

The second type of drag we will cover is *induced drag*. Induced drag is present whenever an airplane is generating lift. Wingtip vortices, discussed later in this chapter, create an upward airflow near the wingtip and a downward flow behind the trailing edge. This downwash causes the vector of lift to incline aft of perpendicular to the wing's relative wind, causing a lift component to the rear of the wing (Figure 1-2). Induced drag is the result of this aft lift component, and this aft component is present when any wing generates lift. Aircraft designers can attempt to reduce induced drag through airfoil and wing design, but this cannot completely eliminate it. Unlike parasite drag, which increases as airspeed increases, induced drag varies inversely as the square of the plane's airspeed. In this case, the slower an airplane is flying, the greater the induced drag it generates. This means that induced drag is highest just above stall speed, such as during landing, and less when the plane is at cruise speeds.

We will discuss lift further later in the next section. For now keep in mind that lift, gravity, drag, and thrust are always present when an airplane flies. Let's move on to a more detailed look at lift.

Induced Drag

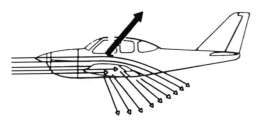

Fig. 1-2 *Induced drag.*

Lift

In this section we are going to take an in-depth look at lift, how it is produced, and how different flight regimes affect the lift the airplane's wings will produce. Figure 1-3 illustrates air flowing over a typical airfoil. As you can see, due to the curvature of the upper surface of the airfoil, the air going over the top of the wing has a longer distance to travel than the air flowing under the wing. According to Bernoulli's principle, if two air molecules start at the leading edge of the wing, and one flows over the top while the other flows under the wing, they will both reach the trailing edge of the wing at the same time. In order for this to take place, the molecule flowing over the top of the wing will move faster than the molecule going under the wing. As a result, and as predicted by Bernoulli's principle, a low-pressure area is created on top of the wing as compared to the pressure under the wing. This is due to the fact that as the speed of the airflow increases, the pressure of the air will drop. So now we have a low-pressure area on the upper surface of the wing and a higher-pressure area on the lower surface of the wing. The higher-pressure area tends to push the wing upward. Pressure differential is one component of the lift created as the plane moves through the air, and is known as pressure differential lift.

Figure 1-4 shows the angle of attack of the wing. By definition, the *angle of attack* is the angle created between the cord line of the wing and the relative wind. The *cord line* is an imaginary straight line that extends from the trailing edge of the wing through the leading edge. The *relative wind* is the direction of the wind opposite the actual direction of movement of the plane. Figure 1-4a has three examples of the angle of attack: low, medium, and high. As the angle of attack increases, the lift generated by the wing also increases, up to a point. A portion of this increased lift results from a further reduction on the pressure on the upper surface of the wing and a greater pressure differential between the upper and lower surfaces. However, as the angle of attack increases, the smooth airflow present at lower angles of attack begins to become turbulent. Beyond a certain angle of attack, known as the critical angle of attack, there is so much turbulence created in the flow of air over the wing that the amount of lift generated is reduced. Figure 1-4a depicts a wing at progressively higher angles of attack, up to the critical angle of attack, and shows the rough flow of air over the top of it and how it increases as the angle of

Fig. 1-3 *Lift.*

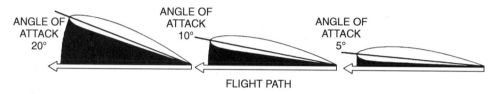

Fig. 1-4 *Angle of attack.*

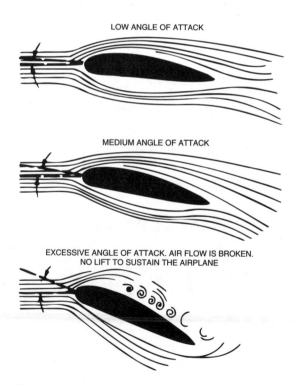

Fig. 1-4A *Angle of attack (low, medium, high).*

attack increases. For most general-aviation aircraft, the critical angle of attack is between approximately 15 and 20 degrees angle of attack.

Wings do not stall all at once when a stall takes place. Instead, the turbulent flow of air works its way across the wing, creating a situation where areas of the wing are producing lift and have a relatively smooth airflow over them, while others have greater turbulence and are not producing lift. Aircraft engineers can pick a wing shape, or planform, that results in stall characteristics best suited for the plane. Figure 1-5 shows six different

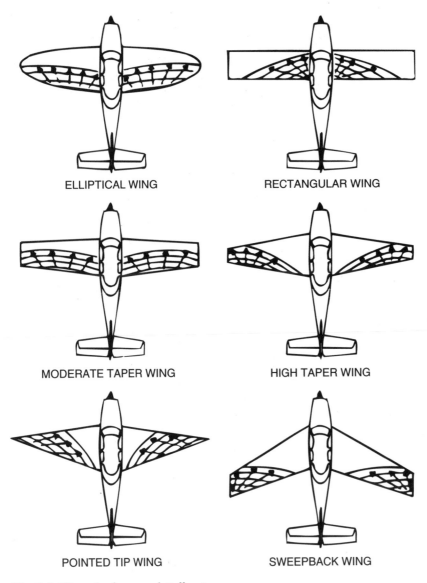

ELLIPTICAL WING

RECTANGULAR WING

MODERATE TAPER WING

HIGH TAPER WING

POINTED TIP WING

SWEEPBACK WING

Fig. 1-5 *Wing planforms and stall patterns.*

wing planforms and the pattern of stall progression on each of them. Notice that the stall normally begins on the aft portion of the wing and moves forward toward the wing's leading edge. On some planforms, the stall starts at the inboard areas of the wing near the fuselage and moves outward. On others, it begins near the wing tips and moves inward.

Manufacturers of light general-aviation aircraft normally try to design the wing so that the ailerons remain effective as the wing stalls. This results in greater aileron control authority during stalls. As you will see in Chapter 3, you should keep use of ailerons to a minimum during stalls, and instead use rudder to maintain wings-level flight. For the most part, general-aviation planes are designed with mild stall characteristics in mind. Anyone who has done stalls in a Cessna 152 can attest to the fact that under most conditions, when the plane stalls there is very little break of the nose downward, or tendency to roll off on a wing, when the plane is flown correctly. These stall characteristics are due in part to the shape of the wing, or its planform.

A plane can reach the critical angle of attack even in level flight. Figure 1-6 shows how a gust of air, in this case an updraft, can cause the relative wind to reach such an angle that the critical angle of attack is exceeded. Anyone who has been flying along on a rough summer day and heard the stall warning horn go off as they fly through turbulence has experienced this effect. In most cases the updraft from this turbulence is so short in duration that the plane quickly recovers with no action from the pilot. However, if you should encounter this type of situation during takeoff or landing, the plane may stall to the point where recovery may be difficult. When flying on gusty or turbulent days, it may be a good idea to maintain extra airspeed as a buffer against stalling due to this rough air.

Now let's take a look at two other types of lift an airplane generates: impact lift and Newtonian lift. Both contribute to the total lift generated by the airplane as it flies. *Impact lift* is the lift the airplane's wings and fuselage create as the air it is flying through strikes their underside. Figure 1-7 gives you an idea of how the airflow presses upward against the bottom of the fuselage and wings. As the angle of attack increases, so does the amount of lift generated by impact lift. If you have ever put your flat palm out the wind of a moving car and angled the forward part of your hand upward, you probably noticed that your hand rose. This is due to the impact of air under your hand forcing it up. When you angle your hand down, the impact of air causes it to descend. This is a very practical example of impact lift and holds true for an airplane's wings and fuselage. I have read that the F-14 Tomcat, a premiere fighter for the U.S. Navy, can generate in excess of 40% of its lift from impact lift at high angles of attack! Given the amount of area the fuselage and wings have, this helps the airplane fly at higher angles of attack.

Newtonian lift is the action/reaction lift that is generated by the wing. Figure 1-8 shows how air flowing off the trailing edge of the wing takes a downward angle. As you might recall, Newton stated that for every action there is an equal and opposite reaction. This downward movement causes the wing to move in the opposite direction it is moving—up. As a result, this also creates a lift component as the airplane moves through the air. The greater the downward force generated by the airflow, the stronger the upward reaction it results in.

As we previously stated in this chapter, when lift is generated by the wings of an airplane, there is a lower-pressure area located above the wing and a higher-pressure area

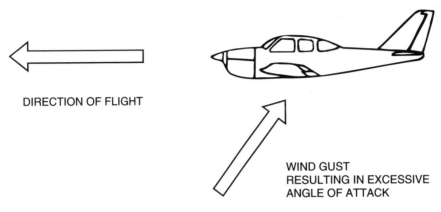

DIRECTION OF FLIGHT

WIND GUST
RESULTING IN EXCESSIVE
ANGLE OF ATTACK

Fig. 1-6 *Critical angle of attack from upgust.*

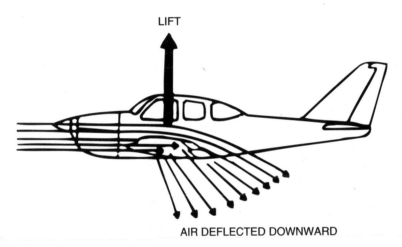

LIFT

AIR DEFLECTED DOWNWARD

Fig. 1-7 *Impact lift.*

below it. Due to this pressure differential, there is a tendency for the higher-pressure air to "roll" over the wingtip to the lower pressure area above it. This rolling air generates a vortex of air that trails behind and below the aircraft, looking and acting much like a horizontal tornado (Figure 1-9). The larger the aircraft, the stronger the vortex it generates. The greatest vortexes are created by large aircraft at slow airspeed, such as airliners during takeoff and landing. It is quite common for pilots to be cautioned by the tower at controlled airports about wake turbulence from large aircraft. The strength of vortexes can be so strong it can roll an aircraft upside down, a bad situation for almost any general-aviation aircraft close to the ground. There have been documented accidents caused by an airplane encountering wake turbulence during which control of the airplane was lost. This loss of control often results in the airplane rolling in the direction of the rotation of the vortex. I have seen footage of tests done where a Boeing 737 was flying into the vortex of another airliner. Even a plane this large was buffeted by the vortex. Smaller aircraft are even more susceptible to the effects of wake turbulence. If you should find yourself in this unfortunate position, and the plane is rolling past ninety degrees, keep in mind the fastest route to upright may be to continue the roll. If the plane does not have the aileron authority to overcome the rolling tendency, keep the plane rolling in the direction of the vortex until you get back to upright. If the plane does go inverted, do not make the

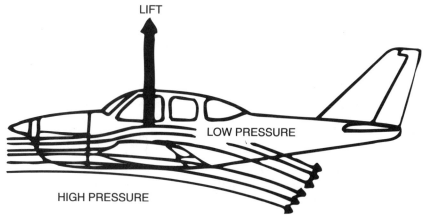

Fig. 1-8 *Newtonian lift.*

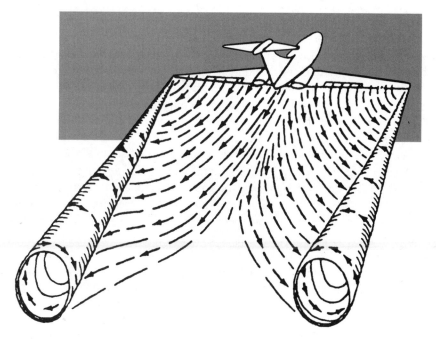

Fig. 1-9 *Wingtip vortex.*

mistake that some pilots do in thinking they can pull the nose down and do a half loop to upright. Unless you have a great deal of altitude, this will probably be the wrong choice for recovery. Figure 1-10 illustrates why the half loop can result in a great deal of altitude loss as compared to a half roll to upright. Another danger is the airspeed you will gain in the dive and the g forces you will impose on the plane trying to pull it out of the dive you are putting it in. Both of these loads on the plane may cause serious damage to the plane, up to pulling the wings off—a situation that would be difficult to recover from.

AIRFOIL TYPES

So far in this chapter we have discussed quite a number of topics related to basic aerodynamics. What lift is, and how it is produced, has received a great deal of attention so far, and we are going to take things a little further in this section. Now we are going to look at airfoils, including a number of different types you might find on airplanes as you walk around the ramp, and the purpose for each of them. If you have spent any time looking at aircraft very closely, you have noticed that the shape of the airfoil can change significantly from plane to plane. This is due to the design of the plane, its intended purpose, the range of airspeeds it will need to fly at, and a host of other design criteria.

Additionally, each airfoil design results in different handling characteristics for the airplane, something you should be aware of as the pilot. While we cannot cover every airfoil and how it causes a plane to fly, we will cover some general characteristics that these airfoils possess.

Figure 1-11 depicts a very common general-aviation airfoil, the Clark Y airfoil. Notice the flat underside of the wing, with the relatively large curvature along the airfoil's upper surface. Figure 1-11a labels each component used in describing the airfoil. These include the wing's leading edge, the trailing edge, the cord of the wing, and the camber of the underside of the wing. The leading and trailing edge definitions are self-explanatory, one being the forward portion of the airfoil and the other being the rearward edge of the wing. The *cord line* of the wing is an imaginary straight line running from the trailing edge of the wing through the leading edge of the wing. *Camber* refers to the curvature of the upper and lower surfaces of the wing. In this figure you can see the *positive camber*, which is the upper surface's curvature, and the *negative camber*, or lower surface's curve. You will also note the *mean camber*, which is the average between the positive and negative camber.

Figure 1-12 shows six airfoil designs. As you can see, the shape of these airfoils is very different, depending on the application the aircraft will be used for. Early airfoils had a great deal of camber on both the upper and lower surfaces of the wing, allowing flight at slow airspeeds. Given the design of the first aircraft, fabric wings, open seating for the pilot, and low-horsepower engines, being able to fly slowly was very important. However, as the horsepower, thrust, and strength of materials increased, higher airspeeds became possible. As a result, the later airfoil and Clark Y airfoil came into use. Today there are many variations of the basic airfoil. When used on general-aviation planes they allow relatively slow airspeeds for takeoff and landing, yet suitably high cruise speeds. These airfoils also provide normally docile handling characteristics through the range of speeds the plane will fly at.

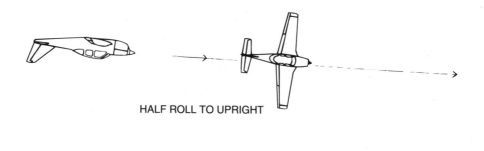

HALF ROLL TO UPRIGHT

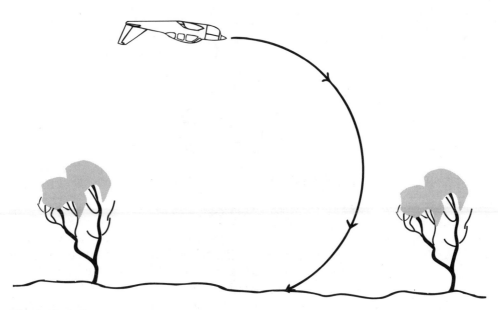

Fig. 1-10 *Roll versus half loop to upright.*

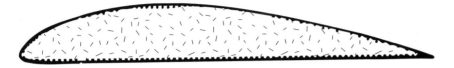

Fig. 1-11 *Clark Y airfoil.*

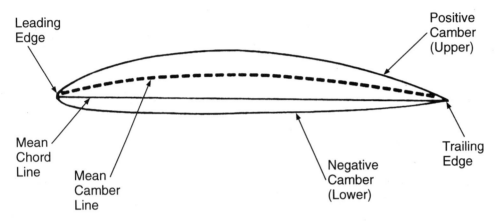

Fig. 1-11A *Airfoil components.*

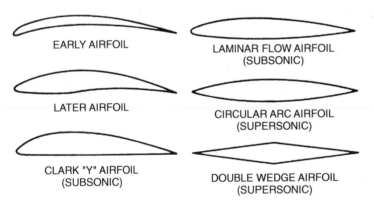

Fig. 1-12 *Airfoils.*

As you look at the airfoils on the right of the illustration, you can see significant changes in the shape. When planes fly at supersonic airspeeds, the wing must be able to generate lift as efficiently as possible, even with the shock wave that supersonic flight causes. The diamond-shaped airfoil on the bottom right of the figure, and variations of it, are popular for supersonic airplanes. However, these airfoils result in less-desirable low-speed characteristics. As you may have guessed, any airfoil selected for use by engineers in an aircraft design is a trade-off. Since most general-aviation aircraft must have a number of different characteristics that are necessary for it to be flown safely and within the intended use of the plane, engineers must pick airfoils that allow that design criteria to be met. However, the plane may not end up with the slowest stalling speed, or highest cruise speed, since most airfoils cannot perform well in both of these different flight regimes. In the next section, we will look at how aeronautical engineers help work around airfoil limitation with lift enhancement devices.

LIFT ENHANCEMENT DEVICES

Airplanes operate at airspeeds ranging from just above stall speed, to cruise speeds, depending on what flight regime they are in. As we have already indicated, every aircraft design is a series of compromises intended to allow the plane to operate best for a given purpose, and then as well as possible in others. For instance, your typical Cessna 172 is built to carry four passengers, fuel, and a little baggage from point A to point B. It is not particularly fast at cruise speeds, but it is very easy to fly, is generally forgiving of minor pilot errors during takeoff and landing, and can land at slow airspeeds to make use of almost any runway. Contrast this with a higher-performance airplane, such as a Mooney. They are faster at cruise speeds but are also more demanding of the pilot during landings due to higher speeds there, as well. Aircraft designers would like to be able to build a plane that cruises at very high speeds and lands at very low speeds, but these are difficult characteristics to build into the same plane economically.

However, they are able to achieve this goal to some degree with the use of lift enhancement devices such as flaps, slats, and leading edge slots. As you will see, each of these devices allows the wing to fly at slower airspeeds without stalling, yet still fly at acceptable cruise airspeeds. We will look at each of these devices in this section. Let's begin with flaps.

Flaps

Earlier in this chapter we discussed camber, or the curvature of the wing's surfaces. Within limits, the greater the camber the wing has, the greater the lift it is capable of producing. Refer back to Figure 1-12 and notice the early airfoil design, then compare that to the Clark Y airfoil. Notice the large camber on the early airfoil's lower surface and how flat the Clark Y's lower surface is. Early aircraft were not able to achieve higher airspeeds, so their designers were interested in being able to generate sufficient lift at low airspeeds, resulting in the highly cambered wing. But today cruise speeds are important to designers, and less cambered airfoils are used to allow those higher airspeeds at cruise.

Airplanes still need to be able to land at relatively slow airspeeds, though, which means the designers needed to find a way to lower the stall speed of the plane. One way they have done this is with flaps. Figure 1-13 depicts three different flap designs: the plain flap, the split flap, and the fowler flap. While there are many different flap designs, we will focus on these three.

As you look at the figure you will see that each of the flap designs increases the camber of the airfoil when the flaps are actuated. As a result, the airplane can fly at slower

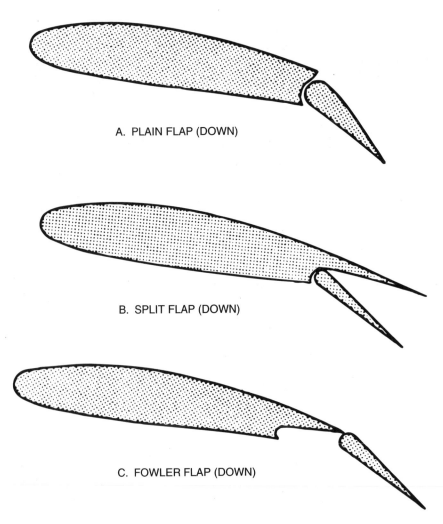

A. PLAIN FLAP (DOWN)

B. SPLIT FLAP (DOWN)

C. FOWLER FLAP (DOWN)

Fig. 1-13 *Plain, split, and Fowler flap types.*

airspeeds without stalling than is possible when the flaps are retracted. Different designs have a wide range of effects on stall speeds. Some planes may only get two- to five-knot decreases in stall speed when flaps are used, while others allow them to fly fifteen to twenty knots slower. Specialized slow-flight aircraft can seemingly hang suspended motionless when their flaps are deployed and land in less than a few hundred feet of runway. I once watched a plane designed to haul cargo in and out of short strips demonstrate its short-field landing ability by landing across the width of the runway, in this case less than 150 feet!

The plain flap is hinged at the rear of the wing and basically pivots down around its hinge point. These are usually controlled either mechanically by the pilot pulling a lever that is directly attached to the flaps or by an electric motor. Flap increments usually come in set increments such as 10, 20, 30, and 40 degrees of flap extension. Depending on the plane, you may get three or four notches of flap increments. While reducing the stall speed of the plane, flaps also generate additional drag as they hang out into the airflow. As a result, a plane can make a steeper approach to landing without the increase in airspeed you would experience without flaps. This can be very useful when you must clear an obstacle such as trees or a building at the approach end of the runway, yet still lose altitude and land. Without flaps it would be necessary to dive to the runway and cause an increase in airspeed. This would then result in longer landing rolls, greater tire and brake wear, and more frazzled pilot nerves. In short- or soft-field landing situations, flaps can reduce the amount of runway used by a significant amount.

The second flap design, split flaps, also increases the camber of the underside of the airfoil, but is located on the aft, bottom portion of the wing. Unlike the plain and fowler flaps, it does not change the camber of the upper surface of the wing. I have seen this flap used most frequently on some general-aviation, light, twin-engine aircraft. This may be to some degree due to the difficulty and expense in designing flaps that fit around the engine nacelles on each wing.

The fowler flaps are the most efficient of the three shown in Figure 1-13. In addition to moving down, they also move back from the trailing edge of the wing. This not only increases the camber of the airfoil on the upper and lower surface, but also increases wing area. Additionally, depending on how they are designed, slots between the wing and sections of flap can improve the flow of air over the wing at high angles of attack, further reducing the speed at which turbulence becomes strong enough to stall the wing. Fowler flaps are popular on some Cessna single-engine planes, and on airliners as well. The next time you are in an airliner, notice how the flaps extend down and away from the trailing edge of the wing during takeoff or landing. This is part of the system that allows an airplane that cruises at Mach .8 to land at relatively slow airspeeds.

Slots

Leading edge slots are not seen on many general-aviation aircraft today but were more popular in aircraft build in the 1940s and 1950s. Often called leading edge slots, a slot is located near the leading edge of the wing, normally toward the outboard half of it. It is comprised of an open slot aft of the leading edge, running lengthwise, parallel to the spar. The first airplanes I had exposure to with slots were Stinsons and Globe Swifts. Figure 1-14 demonstrates the difference between the airflow over the wing of a slotted airfoil at high angle of attack, as compared to that of a wing without slots. As you can see, air is able to flow through the slot, then along the upper surface of the wing. This allows less turbulent airflow to adhere to the wing's upper surface, while the wing without the slot suffers from a loss of lift due to this turbulence. When the wing is at lower angles of attack, the slot has less effect on the flow of air. Slots were often placed in front of the area of the wing the ailerons are attached to, allowing better aileron control at high angles of attack.

Some aircraft in production today are manufactured with slots, but these are normally special-purpose planes designed for short-field takeoffs and landings. Combined with the proper airfoil and flap design, slots can help these airplanes fly at airspeeds that

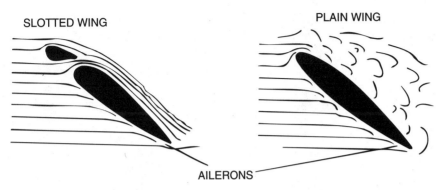

SLOTTED WING

PLAIN WING

AILERONS

Fig. 1-14 *Leading edge slots.*

are surprisingly low. If you do fly an airplane with slots, be sure to check the operations manual to determine their effects on the operation of the plane.

Slats

Leading edge slats are yet another lift enhancement device installed on some aircraft to allow them to fly at slower airspeeds. Like slots, slats are found on very few general-aviation planes. Slats work by extending the leading edge downward, and forward, much like flaps work on the trailing edge of the wing. Figure 1-15 depicts a leading edge flap, which is a variation of the slat. The first leading edge slat I saw was attached to an F-86 Sabre Jet. Slats normally work on slides and rollers, allowing the leading edge of the wing to move down and out from the wing. Figure 1-16 shows a cross-section view of the difference in the camber of the wing with slats extended and retracted. As you can see, the curvature of the wing is increased a great deal, allowing better lift generation at slow airspeeds. Some slats are designed to create a slot between the slat and the wing when the slat is extended. This provides additional ability for the wing to produce lift at higher angles of attack. Leading edge slats can be extended mechanically with motors or hydraulics, or by the weight of the slat against the slipstream. In the case of the latter, the force of air pushing against the slat holds it in a streamlined position. But when the plane slows sufficiently, the weight of the slat forces it into the extended position. If you look at airliners, or military aircraft, you will find that leading edge slats and flaps are often designed into the wing. When combined with fowler flaps, leading edge slats can dramatically increase the camber of the airfoil. This is yet another reason large airliners can land at such low airspeeds.

The next time you are on a jet airliner preparing for takeoff or landing, notice how the shape of the airfoil changes as the leading edge slats and flaps are extended. This allows the plane to take off or land at slower airspeeds and use less runway in the process. The lift enhancement devices we discussed also allow better controllability of the airplane at these slow airspeeds, improving safety as well. When you fly an airplane with any lift enhancement devices, you should thoroughly understand how to use them and how they will affect the flight characteristics of the plane. You should also be aware that some accidents are caused by pilots attempting to fly an airplane equipped with these devices, but failing to use them. More often than not, the pilot attempts to take off or land at the normal airspeeds, then stalls because the lift enhancement devices are not deployed. Airliners are equipped with warning devices that are supposed to let pilots know if they have not properly extended flaps or other devices during takeoff or landing. This mistake can affect even small, general-aviation aircraft if conditions are right. For instance, if you are attempting a soft- or short-field takeoff and do not use the correct amount of flaps as recommended by the manufacturer, you may not leave the runway, or you may stall shortly after leaving ground effect. As we get into takeoffs and landings in Chapter 6, we will discuss the need to correctly use flaps.

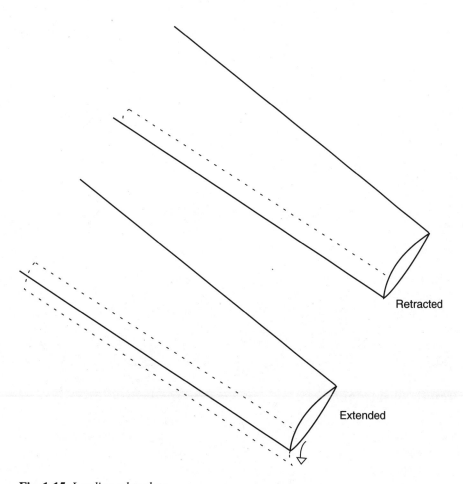

Retracted

Extended

Fig. 1-15 *Leading edge slats.*

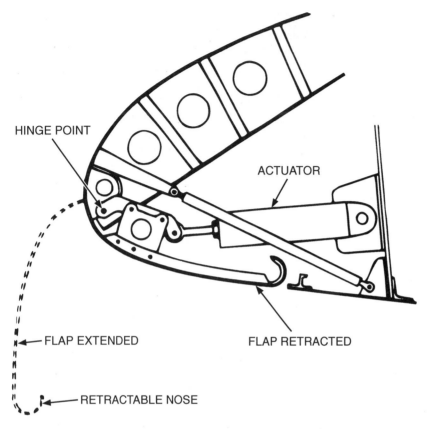

Fig. 1-16 *Leading edge slat extended versus retracted.*

FLIGHT CONTROLS

In this section we are going to discuss the various flight controls most single-engine aircraft are equipped with. There are three primary flight control surfaces: the ailerons, rudder, and elevators. As you will see, these give you control over the three axes that at airplane moves around. We will also continue to discuss flaps on small aircraft, in addition to trim tabs. Before we get into the controls, we are going to define what the movements of an aircraft are, the axis it moves around, and what its center of gravity is.

Aircraft Movement

Aircraft in flight move around three axes: the vertical, longitudinal, and lateral. Figure 1-17 illustrates the three axes of rotation. As you can see, each axis intersects with the other two. This intersection takes place at the airplane's center of gravity, which we will discuss later in this section. The movement about each of these axes of rotation is also labeled in the diagram. *Yaw*, which is the movement of the nose of the plane to the left or right, takes place about the vertical axis. Yaw is be controlled with the rudder via the rudder pedals. Roll takes place around the longitudinal axis and is controlled by the ailerons. Finally, pitch, or the movement of the nose up or down, takes place around the lateral axis; pitch changes are controlled by the elevators. As you will come to understand as you read through this book, understanding the correct use of flight controls—and how control inputs will affect the attitude, airspeed, and control of the aircraft—is one of the basic fundamentals to becoming a safe, competent pilot. Some of the best pilots I know have become so attuned to how an airplane flies, how it feels, what it is telling them in response to control inputs, that they do not consciously think about individual control movements as they fly. Rather, to them pitch, roll, and yaw inputs are like keeping your balance as you walk; you don't think about it, but with each step you adjust the muscles in your body to keep you from falling down. This is the type of pilot that really knows how to fly an airplane. With the proper effort and dedication, pilots can reach a level of competency that will make them much safer as they fly. Knowing how the flight controls work, and how the plane will respond, is part of the foundation you will need to become a safer pilot.

Figure 1-18 illustrates the three primary flight controls of an airplane. These are the ailerons, rudder, and elevator. For most general-aviation airplanes, you will find that this is the standard arrangement. There are some variations, such as the Beechcraft Bonanza,

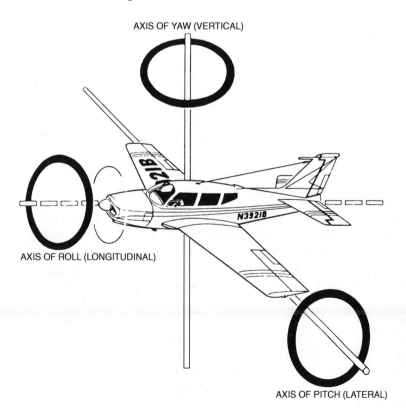

AXIS OF YAW (VERTICAL)

AXIS OF ROLL (LONGITUDINAL)

AXIS OF PITCH (LATERAL)

Fig. 1-17 *Axes.*

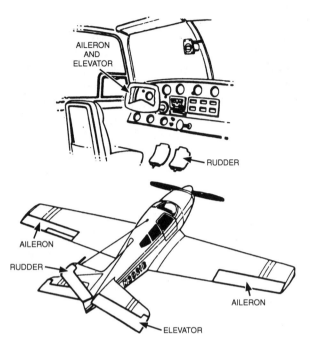

AILERON
AND
ELEVATOR

RUDDER

AILERON

RUDDER

AILERON

ELEVATOR

Fig. 1-18 *Basic flight controls.*

which has a V-tail, and some planes incorporate spoilers on the wings as opposed to ailerons, the Mitsubishi MU-2 being an example, but these are in the minority. As we discussed earlier in this section, the ailerons control roll about the longitudinal axis, elevators control pitch movement around the lateral axis, and rudder controls yaw movement around the vertical axis.

Each of these flight surfaces is controlled by the pilot from inside the cockpit. Ailerons are controlled by the control wheel, or stick in some aircraft, through turning the control wheel, or moving the stick from left to right. To roll the plane to the left, you turn the control wheel to the left; to roll right, you turn the control wheel to the right. If the plane you are flying has a control stick, then lateral movement of the stick to the left or right rolls the plane to the left or right. Figure 1-19 shows how ailerons cause the plane to roll. As you can see, in a roll to the left the right aileron moves down, while the left aileron moves up. We learned earlier in the chapter that greater cambers on the airfoil generates more lift. With ailerons we are essentially increasing the camber of the wing that has the lowered aileron, which increases lift for that wing. The wing with the raised aileron suffers from a reduction in lift, the end result in this example being that the right wing is generating more lift than the left wing. This imbalance in lift generated by the wings causes the right wing to rise and the left wing to drop. As a result, the airplane rolls to the left about the longitudinal axis. The same holds true when you turn the control wheel to the right. The left wing generates more lift than the right, and the plane rolls right.

The elevator is also moved through the control wheel through forward and backward movement. To raise the nose of the plane, you pull back on the control wheel; to lower the nose you move the control wheel forward. Figure 1-20 illustrates how this motion takes place. When you pull back on the control wheel, or control yoke as it is sometimes called, you are changing the camber of the elevators, which are just another set of airfoils. In the first example the elevator pivots upward around the hinge point. This causes the larger camber to be on the upper surface of the elevator, and greater lift is generated on the underside of the elevator. This causes the lift to force the tail of the plane down, pivoting the plane around the center of gravity, and the nose of the plane rises. When you push the control yoke forward, the elevator pivots downward, generating more lift on the top of the elevator surface, which results in the tail of the plane rising. This movement forces the nose of the plane down. For planes equipped with control sticks, the forward and backward movement of the stick has the same effect as forward and backward movement of the control yoke.

The rudder is probably the most misunderstood primary control surface. If you ask pilots what causes an airplane to turn, a surprising percentage will answer the rudder. Unfortunately, this is not correct. While the rudder is used to control yaw, in properly flown coordinated flight, it is not used to turn the airplane. The movement of the rudder is controlled via the rudder pedals, which you place your feet on when you fly. The rudder is another airfoil surface that generates lift through changing the camber. Figure 1-21 shows how movement of the rudder to the left and right generated changes in camber, and lift, just as we saw with the elevators. When you push the left rudder pedal, the rudder pivots to the left, causing more lift on the rudder's right side. This causes the tail to pivot to the right and the nose to move left. When the right rudder pedal is pushed, the rudder moves right; increased lift on the left side of the tail causes it to pivot left, and the nose moves right. Like the other motions, this movement takes place about the plane's center of gravity.

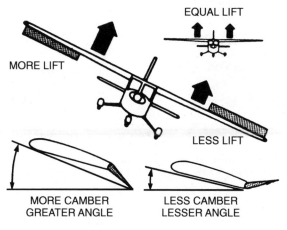

Fig. 1-19 *Lift changes caused by ailerons.*

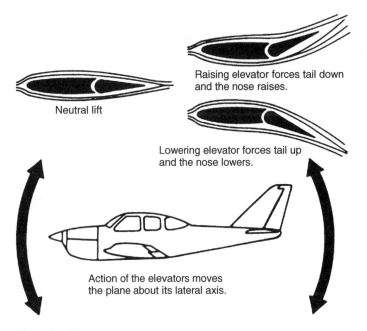

Raising elevator forces tail down and the nose raises.

Neutral lift

Lowering elevator forces tail up and the nose lowers.

Action of the elevators moves the plane about its lateral axis.

Fig. 1-20 *Elevator movement.*

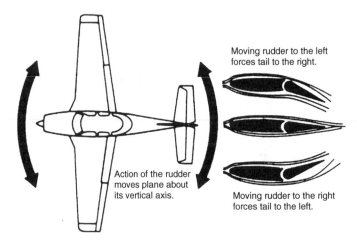

Moving rudder to the left forces tail to the right.

Action of the rudder moves plane about its vertical axis.

Moving rudder to the right forces tail to the left.

Fig. 1-21 *Rudder movement.*

At this point you might be asking yourself, "Okay, then what causes an airplane to turn, and what is the rudder for?" Earlier in this section I mentioned "coordinated flight." When you fly an airplane, you want to use the controls in what is referred to as a coordinated manner. To understand what that is, we are going to look a little deeper into basic aerodynamics, starting with what are known as lift vectors. Please don't be run off by the name; as you read you will see just how simple a concept this is.

LIFT VECTORS

We have already learned that when an airplane is in level flight, lift equals gravity. Figure 1-22 shows a plane in both straight and level flight, and in a turn. In part A of the figure, all lift is exactly opposite the effect of gravity. Portion B of the illustration shows the airplane in a banked turn. Notice how some of the lift that had been opposing the vertical pull of gravity is now working against the force of gravity, while some lift is vectored in the direction the plane is banked. While this is a simple model of a more complex action, the lift that is angled toward the direction of the bank is what actually causes the plane to turn in that direction, while some lift is used to maintain the plane's altitude. This is a very important concept, and you need to understand it to really begin to know how an airplane flies. Notice that the overall lift vector is greater than the lift vector for a plane in level flight. This is due to the fact that some of the lift that was used to hold the plane at altitude is being used to turn the plane, yet the plane must still generate enough lift to oppose the force of gravity. The amount of the vertical lift component needs to remain equal to that needed in level flight, and more lift is being used to turn the plane, so the total overall lift in a turn must be greater when the plane is turning than when it is in level flight.

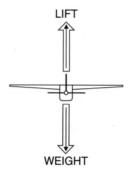

Fig. 1-22A *Lift vector: straight and level.*

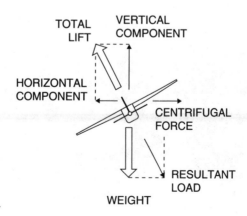

Fig. 1-22B *Lift vector: bank.*

How do we increase lift when in a turn, then, to allow us to maintain altitude and let the airplane turn? We already know that increasing the angle of attack, up to a point, will generate greater lift. When we turn we must pull back slightly on the control yoke, causing an increase in the angle of attack and the total amount of lift. At this point we have two actions needed to turn the airplane and maintain level flight: an aileron input to cause the plane to roll to the desired bank angle, which vectors some lift to turn the plane, and an aft movement of the control yoke to increase the angle of attack and therefore the total lift the wings are generating. So what do we need a rudder for?

When you make the aileron input, one aileron moves down, while the other moves up. As it turns out, the lowered aileron generates more drag than the raised aileron. This asymmetrical drag causes the nose of the plane to yaw in the direction of the lowered aileron. For example, when you turn left the right aileron moves down, causing greater drag on the right wing. This drag causes the nose of the plane to yaw right; this yaw is known as *adverse yaw*. To compensate for adverse yaw rudder is input, which helps prevent the nose from yawing. For a left turn, left rudder is used; in a right turn, right rudder is used. When done correctly, the amount of rudder exactly cancels out adverse yaw from the aileron.

An instrument known as the ball is found on the instrument panel of most aircraft. Figure 1-23 shows the ball, a small black bead in a curved tube filled with liquid. In a coordinated turn the ball stays centered in the glass tube, but in an uncoordinated turn the ball slips to the inside of the turn or skids to the outside of the turn. Parts B and C of the figure show how the ball looks in a slipping or skidding turn, while part A shows the ball centered during the turn. How much rudder will be required to keep the ball centered depends on the airplane, the airspeed it is flying at, the angle of the bank, and a number of other factors. Early in their flight training, students will often stare at the ball during a turn, trying to get the correct amount of rudder input. If a pilots practice, though, they can feel whether a turn is coordinated. This feeling is often called "seat of the pants flying" because pilots can tell by the feeling their rear end has in the seat—and whether it is being pushed to one side or the other. A pilot in tune with the airplane can tell just by the feel it has whether the rudder input is correct and the turn is coordinated.

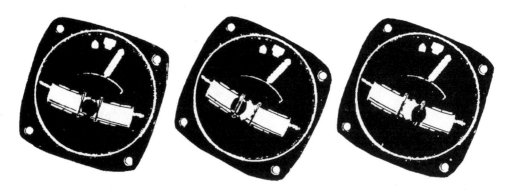

Fig. 1-23 *Ball and needle.*

Now we understand the use of all three primary flight controls in a turn; the ailerons cause the airplane to roll, vectoring lift in the direction of the turn. The elevator is used to increase the lift necessary to maintain a constant altitude during the turn, and the rudder is used to overcome adverse yaw caused by the lowered aileron. As you gain more experience in using the controls correctly, the coordinated use of these controls will become less mechanical, allowing smoother entry and exit from a turn. We will discuss use of the controls, the instruments, and the visual picture you will have outside the airplane in Chapter 2. Let's continue the chapter with a discussion of a secondary flight control, the trim tab.

TRIM TABS

When you are flying, outside forces are constantly influencing the actions of the airplane. Among these influences are attributes associated with the plane itself. How you have it loaded, power settings, and other factors can influence how the airplane behaves during flight. For instance, in a climb, you will find it is necessary to maintain backpressure on the control yoke, while in a descent you may find that a constant forward pressure is necessary. Some planes require that you keep right or left rudder in during level flight to keep the ball centered, depending on how they are rigged. Other airplanes have tendency to roll off to the left or right, depending on fuel and weight distribution. These continuous control inputs can not only be tiring for the pilot, but also a distraction. When the pilot is in a busy environment, remembering to keep a slight amount of right aileron in to keep the airplane from rolling to the left can cause unnecessary distractions. The same is true of constant elevator inputs on long climbs or descents.

To help overcome the tendency of almost every airplane's need for constant control inputs during some phases of flight, aircraft designers have incorporated a device known as a *trim tab* into the flight control surfaces of most general-aviation aircraft. Some, such as in Figure 1-24, are ground adjustable, meaning you must be on the ground and outside the aircraft to change the trim tab's angle. In the case of Figure 1-24, this external trim tab is used to control the rudder and can be adjusted to help keep the ball centered during cruise flight. Others are adjustable from inside the cockpit while in flight. We will begin the discussion with a review of how trim tabs work.

A trim tab acts in the same relationship as an aileron does to the airfoil, using the change in camber and the impact of air from the slipstream to change the position of the control surface. Figure 1-25A shows an elevator trim tab, but the same principles apply to rudder or aileron trim tabs. As you can see, the elevator trim is hinged at the trailing edge of the elevator. A small wheel or knob is located inside the cockpit, within reach of the pilot. As the control wheel is moved, linkage causes the trim tab to pivot about its hinge point, causing the trim tab to deflect the control surface to some degree as a result of the impact of airflow. Figure 1-25B shows the trim tab pivoting upward, resulting in the deflection of the elevator downward. As a result, the nose of the plane will drop, just as if you had pushed forward on the control yoke, forcing the elevator downward. The amount of elevator deflection depends on how much trim tab the pilot inputs, the airspeed of the plane, and other factors. Figure 1-25C shows the trim tab

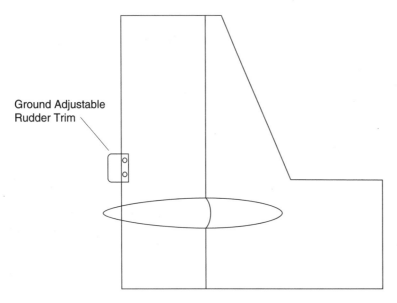

Fig. 1-24 *Ground adjustable rudder trim.*

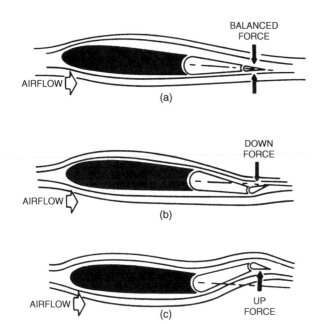

Fig. 1-25 *Elevator trim (a,b,c).*

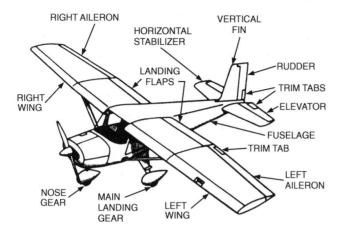

Fig. 1-26 *Aileron and rudder trim.*

forcing the elevator upward, resulting in the nose of the plane rising. You will find that elevator trim is very useful during extended climbs, descents, and also during takeoffs and landings. Trimming the plane for level flight is also necessary, and continued adjustments to the trim will be necessary as fuel loads and airspeeds change.

Figure 1-26 shows both aileron and rudder trim tabs. Most single-engine general-aviation planes are equipped with elevator trim, but a smaller percentage have aileron or rudder trim. Like the elevator trim tabs, aileron and rudder trim cause the control surface to be deflected, and the plane behaves in the same manner as if the pilot had made the control input. As we already discussed, aileron trim can help reduce a rolling tendency a plane may exhibit. This could be due to unequal weight distribution, such as more fuel in one wing tank than another, or more passengers on one side of the plane. Sometimes how a plane is rigged, or the twist that is built into the wings, can also result in a rolling tendency. Rudder trim is used to overcome any yawing tendency a plane may have, such as during a climb (more about those forces in Chapter 3). At this point we will just state that there are flight regimes during which a plane will have a tendency to yaw to the left or right. If you are in a long climb and the plane needs constant right rudder during that climb, it reduces pilot workload to trim the rudder to compensate for this yawing tendency.

CENTER OF GRAVITY

We are going to finish the chapter with a review of center of gravity. Simply put, the *center of gravity* is the point that the airplane you are flying balances around. We've all taken a short length of wood, or a stick, and balanced it lengthwise on our finger. When you find the point where the piece of wood balances on your finger, you have found its center of gravity. Center of gravity is important to airplanes because in order for them to fly safely, the position of the center of gravity must be within a range along the wing. Figure 1-27 shows an airplane and its center of gravity, in addition to the acceptable center of

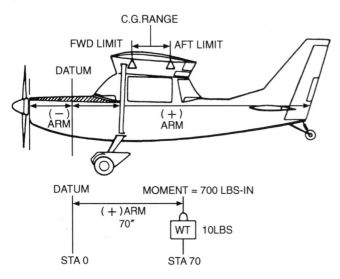

Fig. 1-27 *Center of gravity.*

gravity ranges for that plane. Airplanes have a forward and aft limit to the center of gravity. If the center of gravity lies beyond those limits, it will have an adverse effect on the
flight characteristics of the plane and may make it unsafe to fly.

At this point you may be wondering why the center of gravity changes and how you
find out what the center of gravity's position is. You control the center of gravity by how
you load passengers, fuel, and cargo into the plane. Imagine the stick you had balancing
on your hand, and now you place several coins on one end of the stick. This is going to
cause the end with the additional weight of the coin to drop. In order to balance the stick
again you will need to move your finger to a new point, closer to the weighted end. The
addition of weight to the stick has moved the center of gravity, just as placing passengers,
fuel, and cargo affects the center of gravity of the plane. If the plane is loaded correctly,
the center of gravity remains within the allowable range, and the plane will not be
affected adversely. However, if you load the plane so that the center of gravity falls outside the range, you may not be able to control the plane.

For instance, let's say you are flying a four-place, single-engine airplane and carrying two passengers, plus baggage. If you decide to put both passengers in the rear seats,
and place their baggage in the aft baggage area, you have placed a great deal of weight
toward the rear of the plane. The result could be that the center of gravity will have
moved so far aft that it may fall outside acceptable ranges. In this case you may need to
move one of the passengers to the front seat and redistribute their baggage to more forward areas.

There are several methods used to compute the center of gravity, or weight and balance as it is sometimes referred to. The operations manual for the plane you are flying
will have a weight and balance section, which must be used for the plane you are flying. Every plane is different in its original weights and location of the center of gravity

at its empty weight, so you need to determine the correct center of gravity for the plane you are flying using that airplane's operations manual. It will contain the weight and balance data specific to that plane. We will not get into a detailed discussion regarding computing weight and balance, but you should be aware of the need to know what the limits are for the plane you fly, and how to compute them. The operations manual will provide specific information, and examples of how to compute, the weight and balance for the plane you fly.

CONCLUSION

We have covered a great deal of information in this chapter. We started with a basic discussion of airfoils and how lift is generated. Different airfoil types, Bernoulli's principle, and the various types of lift were discussed in detail. Lift enhancement devices—such as flaps, slats, and slots—were also covered. Each of these devices allows the wing to generate sufficient lift to maintain flight at slower airspeeds than would be possible without them.

The primary flight controls—rudder, ailerons, and elevator—were also reviewed. How each of these control surfaces affects aircraft control and how they are used by the pilot were given a great deal of attention. We also looked at what actually causes an airplane to turn, in addition to how to correctly coordinate the use of all three primary flight controls in a turn. Remember, as we discuss various fight maneuvers in coming chapters, you will need to understand the proper control inputs and how to keep the maneuvers coordinated. Finally, we defined the center of gravity of an airplane and took at brief look at why it is important to you when you fly.

This chapter has covered the basics you will need to know and understand throughout the rest of the book. In order to fly well, having this foundation is crucial to being able to do more than just drive a plane from point A to B. In everyday flying, knowing this information can make you a safer pilot, not only when a true emergency situation exists, but also in avoiding potentially serious problems. If you have questions about what was covered in the chapter, find a qualified flight instructor who can answer them for you. Most accidents are due to pilot error. With a better understanding of how to fly, some of those accidents might be avoidable.

2
Basic Flight Maneuvers

In Chapter 1 we discussed some very fundamental concepts relating to aerodynamics. In Chapter 2 we are going to build on that information and begin to apply it to flying the airplane. Like any activity, learning the basic skills allows you to master the simple tasks, then move on to more complex areas. Learning to fly an airplane is based on this concept. We learn the basic flight maneuvers, then build on those skills, learning more difficult maneuvers.

This chapter will discuss a number of those basic skills, which include straight and level flight, climbs, descents, turns, and steep turns. We will cover the control inputs for each maneuver, the power settings that can be used, how the instruments will look, and what the view from the airplane will be. When you finish reading this chapter you should come away with an understanding of how each maneuver is flown and the mistakes that are commonly made by pilots. In Chapter 1 you learned what the primary flight controls are used for and how to use them in a coordinated manner. As we discuss each of the maneuvers in this chapter, we will also look at how to fly them in a coordinated manner. Before we get into flying some of the basic maneuvers, let's start with a review of some of the basic instruments you will need to be familiar with.

FLIGHT INSTRUMENTS

When you learn to fly, you will not only need to learn how to fly each of the maneuvers by using controls, but also through looking outside the airplane for attitude reference, and inside the plane at its instrumentation. In this section we are going to cover the basic flight instruments that will be found in most single-engine general-aviation aircraft. As you fly, you undoubtedly will notice that the instrumentation in different planes can vary greatly. Some have a minimal number of instruments, while others have every electronic gadget known to pilots. We will not be able to cover every conceivable instrument, but we will concentrate on a basic panel consisting of the airspeed indicator, the altimeter, the artificial horizon, vertical speed indicator, and the directional gyro. In most airplanes these are grouped in a "T" on the panel, and are referenced as part of the scan you use while you are flying.

Figure 2-1 shows a basic panel layout. As you can see, the panel contains a number of instruments. Each has a piece of information that is important to you as you fly. Let's begin with the center of the panel. At the top center is the artificial horizon. This is a gyroscopically driven instrument. A vacuum pump attached to the engine causes a flow of air across the instrument, which spins the gyroscope in it up to speed. Once the gyroscope is at speed, it will attempt to maintain a constant orientation as the plane moves, giving it the ability to mimic the horizon's constant position. Figure 2-2A shows the artificial horizon when the plane is in level flight. As you can see, the small airplane in the instrument's foreground represents the plane you are flying, while the background represents the horizon, earth and sky, outside the airplane. As the plane moves, the background of the artificial horizon moves. When viewed in relation to the small airplane in the instrument, you can maintain a reference to the horizon, even if the outside horizon is obscured. Figure 2-2B shows the artificial horizon when the plane is in a climbing turn to the left, while 2-2C shows the plane in a descending turn to the left. Please note that the small plane in the instrument does not move; the horizon behind it moves.

The instrument located directly below the artificial horizon is the *directional gyro*, also know as the DG. Like the artificial horizon, it is a gyroscopic instrument driven by the suction pump on most single-engine aircraft. Figure 2-3 shows the directional gyro. The instrument acts much like a compass but does not use the magnetic field of the earth for direction-finding purposes. Instead, as part of the runup prior to takeoff, you set the

Fig. 2-1 *Basic panel layout.*

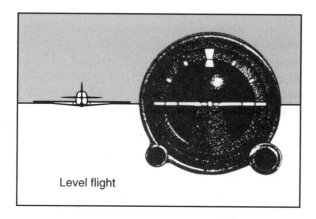

Fig. 2-2A *Artificial horizon: straight and level.*

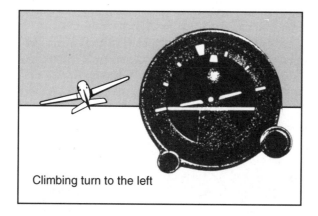

Fig. 2-2B *Artificial horizon: ascending turn to left.*

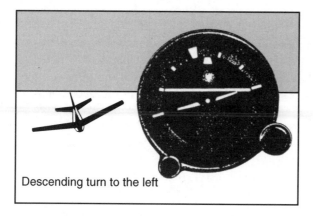

Fig. 2-2C *Artificial horizon: descending turn to left.*

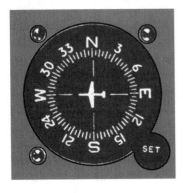

Fig. 2-3 *Directional gryo.*

DG to the same heading the compass shows. The gyro in the instrument then maintains this alignment and indicates the plane's heading. As you turn right or left the DG reflects the airplane's new heading. A word of warning concerning the DG, though. After a period of time a gyroscope precesses, or changes, its alignment. The directional gyro in most airplanes has some tendency to drift off the correct heading as time passes. This can mean that as you fly along, the heading shown on the DG changes in comparison to the one shown by the magnetic compass. Over an extended period of time this drift can cause you to end up off course, and in extreme cases so far off that you may have trouble figuring out where you are. Every fifteen minutes or so you should compare the compass heading to the DG heading and reset the DG to match the compass when necessary. For planes with older instruments, or those with failing suction pumps, you may find that the DG will not hold the correct heading for any period of time. When it comes to low vacuum pressure, the same is true of the artificial horizon. I have had flights where low vacuum pressure from the pump caused the artificial horizon to give erroneous readings, indicating a bank when the plane was actually flying straight and level. In heavy instrument conditions this can cause the pilot some distress, but with a proper scan you should be able to quickly determine that an instrument is failing, which instrument it is, and then continue to fly using the remaining instrument.

The airspeed indicator is the next instrument we will cover. Like the speedometer of a car, the airspeed indicator shows how fast the airplane is moving. Figure 2-4 shows the airspeed indicator. As you can see, there is a needle that move clockwise to show the airspeed. The airspeed indicator receives its readings as a result of pressure differentials caused by air entering a probe called a *pitot tube* and static air pressure from the static vent. The difference in these pressures causes the airspeed indicator to respond, indicating the plane's airspeed. As you look at the airspeed indicator you will note that there are several color bands on it: white, green, yellow, and red. Each of these bands represents a range of airspeeds for the plane. The white band is the flap range of airspeeds. The high end is the maximum airspeed the flaps may be extended at, while the lower end of it is the stall speed of the plane with flaps fully extended. If flaps are kept out above the airspeed indicated at the top of the white arc, you could damage the flaps and other structures of the wing. In a worst-case situation, it is possible you could lose the flap entirely as the forces from the airstream striking the flap place so much load on it that the flap is ripped from the wing.

The green arc is the range of airspeeds that the plane should fly under normal conditions. The bottom of the green arc is the stalling speed of the plane with flaps retracted, or in the clean configuration. As you can see from the figure, the stalling speed is higher when flaps are not used, which fits with our discussion of flaps earlier. At the top of the green arc is the airspeed that should be used for normal cruise, although many planes are not able to cruise in level flight at this airspeed but will fly at a slower airspeed.

A yellow arc starts at the end of the green arc, and this is the range of airspeeds the plane should be flown in only with caution. At these airspeeds it is possible to damage the airplane if severe turbulence is encountered or rapid control inputs are made. For this reason, you should avoid flying at airspeeds in the yellow arc except in those cases where you know the air is very smooth, and even then fly with caution.

At the top of the yellow arc is a red line, the maximum airspeed you can fly at. Above this airspeed the airplane can be structurally damaged due to the loads that you place on the plane. Additionally, control surfaces can begin to flutter above this airspeed, causing difficulty in control of the airplane and possible structural damage. In some cases flutter becomes so pronounced that the control surface can be ripped from the plane, or the entire aircraft can be torn apart from the vibrations. We will discuss other airspeeds you need to know about in Chapter 3, but when you fly remember that the airspeed indicator provides a great deal of information about the plane you are flying.

The altimeter is the next instrument we will focus on. Figure 2-5 shows an altimeter, which operates by measuring changes in barometric pressure. An altimeter indicates the plane's altitude above an atmospheric pressure level by sensing the atmospheric pressure that the plane you are flying is at. Altimeters normally have three needles: one showing feet, one showing hundreds of feet, and finally one showing thousands of feet. Most altimeters also have a small window that is used to set the barometric pressure for your location. This then allows the altimeter to indicate the correct altitude.

The last instrument we will discuss is the vertical speed indicator. Like the altimeter, the vertical speed indicator is a pressure-sensing device that is used to indicate the

Fig. 2-4 *Airspeed indicator.*

Fig. 2-5 *Altimeter.*

rate of change in your airplane's altitude. Another way of saying this is that the vertical speed indicator shows how rapidly the plane is climbing or descending. This is usually expressed in hundreds of feet per minutes. You will find that single-engine, general-aviation planes normally have climb rates of around 500 to 1000 feet per minute, depending on a number of factors.

As you fly, especially in Visual Flight Rule (VFR) conditions, you will want to get into the habit of scanning the instruments and looking outside the airplane as you fly. Many new students fixate on the instruments, ignoring the need to look outside the airplane as they fly. For private pilots flying in VFR weather, the instruments in the panel should be a cross-check to what you see outside the plane. On hazy days, or at night, the instruments can be very useful in making sure you are flying straight and level, maintaining your altitude, and flying on the correct heading. Avoid relying on them to the point of not scanning outside the aircraft, though. When you are flying you need to look for other aircraft in your vicinity, to check landmarks along your route of flight, and to just simply enjoy the view. Too many pilots have become so reliant on instruments that they no longer are comfortable flying on a very basic instrument panel. While using the instruments correctly will allow you to fly more safely and accurately, you should also be able to fly just as well if an instrument should fail in flight. As we cover some of the basic flight maneuvers in the remainder of the chapter, we will compare what you will see looking outside the plane with what the instruments will indicate as well.

STRAIGHT AND LEVEL

While it may seem like a simple activity, holding a plane in straight and level flight requires that the pilot understand how to use the proper power settings, what different power settings will cause the airplane to do, and how to trim the plane to make it easier to maintain level flight. Additionally, turbulence can cause the plane to gain or lose altitude, or to stray from wings-level flight. How you react to these external inputs, and use the controls to correct back to level flight, affects how the airplane flies and how smooth the ride is for your passengers. You may be thinking, "All I need to do is hold the control yoke still, and things should work out." While once in a great while you get a glass-smooth day, and you can trim the plane up for straight and level and it stays there. But the rest of the time you must actively work to maintain your altitude and keep the wings level.

To fly a plane straight and level—which means maintaining your altitude at a constant value, holding the heading you want to fly, and keeping the wings level with the horizon—you must know the "sight picture" you should be looking for, how to use the controls to achieve that attitude, and what to do to correct for variations. We will discuss climbs later in this chapter, but as an airplane reaches cruising altitude, you will begin to level off by reducing the amount of up elevator that you have input by reducing back-pressure on the control yoke. This should be a smooth, gradual release, and not an abrupt shove forward. As the nose of the plane drops, you will see the attitude indicator, as well as the horizon outside the airplane, reflect that change. The correct position for the control yoke will vary from plane to plane, and even from flight to flight in the same plane, depending on how it is loaded.

As you lower the nose to the level flight attitude, you should have an idea of where the horizon will be in relation to the nose. For most aircraft, the nose will be below the line of the horizon during level flight. After you gain experience with a plane, you can normally lower the nose to just about the right position in relation to the horizon just from experience. Once you have the nose at the right position, you should check the altimeter and vertical speed indicator to determine if you are climbing, descending, or maintaining the correct altitude. The vertical speed indicator can be very useful in figuring out if there are small corrections that need to be made.

If you are transitioning to straight and level from a climb, you will probably need to pull back the engine power from its climb setting to the cruise setting you want to use. As you reduce power it is likely the nose of the plane will want to drop slightly, and you may need to readjust the control yoke's position to maintain the correct altitude. Power settings can have a tremendous effect on how the airplane will fly. For instance, if you are flying straight and level and you increase the power without changing any control inputs, in most cases the plane picks up more airspeed. This greater airspeed results in more lift, and the airplane has a tendency to climb. To maintain the same altitude after a power increase, you will need to push forward on the control yoke, which keeps the airplane from climbing. The amount you will push forward depends on how much power was added and the characteristics of the plane you are flying. You should also retrim the elevator, setting it to reduce the pressure you are holding on the control yoke to neutral.

We covered trim tabs in the previous chapter. Normally the trim control is a wheel or a crank handle that you turn to change the trim setting. For elevators they are usually marked "nose up" or "nose down" for the direction you turn the control. In this case you will make a minor "nose down" trim change to help reduce the pressure you are holding on the control yoke. You will find that small trim inputs are usually all that are necessary to make these minor trim-setting changes. You should also notice the airspeed indicator creeping upwards. Eventually the plane will stabilize, with lift equal to gravity, and drag equal to thrust.

If you decrease power, the opposite is true if you want to maintain a constant altitude. You will need to pull back slightly on the elevator to hold the nose at the correct attitude. You should also retrim the elevator to reduce the pressure you are holding to neutral. The airspeed will slowly bleed off, and the plane will stabilize. Of course, there is a limit to how much you can slow the airplane and still hold altitude. We discussed stalls to some extent in Chapter 1, and if you keep slowing a plane down eventually it will stall.

Once you have the airplane trimmed for level flight you can begin to verify that you are holding straight and level. At this point you have scanned outside the plane to make sure your attitude is correct, and verified that with the artificial horizon. At the same time you should verify that the wings are level. If the plane is banked left or right, the horizon will appear banked as you look outside of the airplane. The artificial horizon will also show that you are banked. You can then use the compass and directional gyro to confirm your heading.

As you fly along in straight and level flight, the plane will be occasionally influenced by the air you are flying through. Wind gusts, turbulence, updrafts, and down drafts all affect the airplane as it flies. You will find that you need to continuously make minor inputs to correct for these outside disturbances to straight and level flight. If the wings are rocked to the left or right, you can use aileron to get them back to level. The same holds true for nose up or down. You can be flying along and encounter an updraft, which wants to push the airplane upward. To counteract the effects of the updraft you will need to push the nose down. As you fly out of the updraft you will need to release the forward pressure to avoid losing altitude. Here's a tip: If you are flying along on a day where the wings are getting rocked a fair amount by turbulence, try using the rudder—as opposed to the ailerons—to pick up the low wing. For small disturbances of wings' level, using the rudder is an effective way to get them back to level, and it's easier on the passengers. The movements are less abrupt and have fewer tendencies to cause motion sickness in a passenger. For instance, if the left wing drops, use right rudder to lift it. As the wing returns to level, go back to neutral elevator. On some turbulent days your feet are doing a continuous dance on the rudder pedals to keep the wings level.

Once you gain experience in holding a plane at straight and level, you will find it is not nearly so mechanical as the previous description might indicate. Flying the airplane is a fluid activity, with you making constant, small inputs to hold the plane's wings level and the nose at the correct attitude. You should also continuously scan both the instruments inside and outside the plane. Don't become complacent and fix your attention to one thing. Unfortunately, there are many pilots who are in the habit of doing just that. Once the plane is set up, they let the plane drive itself along and miss many of the cues that tell them what the airplane is doing. Keep your senses open as you fly, and you will find that the plane is constantly giving you feedback.

CLIMBS AND DESCENTS

We briefly mentioned climbs in the last section, but at this point we are going to take a more detailed look at both climbs and descents. Both of these maneuvers are an integral part of flying but are often misunderstood in how to execute them properly. When flown properly, a pilot can level off from a climb or descent in a smooth transition to straight and level. As we get into landings in Chapter 6 you will see that a constant, well-planned descent is important to making consistently good landings, and a well-planned climb altitude can help reduce your time in flight and save fuel. Let's begin with climbs.

As we already covered, climbs are necessary to gain the altitude you need to fly. During a climb the plane is generating more lift than the force of gravity, which allows it to gain altitude. As we covered in Chapter 1, this additional lift can be generated through increasing the angle of attack or increasing speed. Both of these actions result in greater lift generation. In practice, you will find that when you want to climb, you will increase engine power, which results in an increase in airspeed, and increase the angle of attack through backpressure on the control yoke.

For instance, let's say we're cruising along at 3000 feet above ground level (AGL), and we are asked by air traffic control to climb to 5000 feet. To initiate the climb you will do two things: First, add power as recommended by the aircraft's manufacturer. This will vary from plane to plane, but for many aircraft, climb settings normally are full power. Then increase backpressure on the control yoke, raising the nose of the plane. There are two airspeeds that you will be concerned with during climbs: the best angle of climb airspeed and the best rate of climb airspeed.

The best rate of climb, expressed as V_y, gives you the greatest gain in altitude over a period of time. The best angle of climb airspeed, known as V_x, results in the greatest altitude gain in a given distance. Generally, for takeoffs you will initially climb at V_x airspeeds for that plane. For climbs from a given altitude, such as in our example, many aircraft manufacturers recommend using the V_y airspeed. For most airplanes, V_x is a lower airspeed than V_y. As a result, the angle of attack is greater at V_x than it is at V_y, resulting in the nose of the plane being higher. For extended climbs, the use of V_y keeps the nose of the plane lower and allows better airflow into the engine compartment. This provides better cooling to the engine and reduces engine wear. We will discuss V speeds and their definitions further in Chapter 6.

Back to our example, once you have increased power you will raise the nose. The amount of nose up you will use depends on the plane you are flying, the atmospheric conditions, and other factors. Normally, with a little experience, you can estimate the amount of up elevator you will add. As you add the nose-up input, the airspeed will begin to fall. You are now taking some of the additional power and lift and converting it to an altitude gain. Normally you will want the airspeed to stabilize at the correct climb airspeed, whether that is V_x, V_y, or another airspeed recommended by the manufacturer. Once the airspeed has stabilized you can retrim the elevator to reduce backpressure on the control yoke to neutral. It is a good idea to retrim the plane when in a climb for two reasons: it reduces the pilot's fatigue, both from a physical and a psychological aspect. If you don't have to hold the control yoke back using your arm, you will have an easier time holding a constant climb attitude for any length of time. Not having to concentrate on holding the elevator back also allows you to more easily monitor what is going on, as opposed to concentrating on the control input. Many new pilots do not realize how important trimming the airplane correctly can be, and they get into the habit of "muscling" the airplane around. They then end up focusing so intently on holding the control yoke, or keeping the airspeed steady, that they forget to monitor heading, altitude, and other instruments and conditions.

As you approach the altitude you are climbing to, you will need to gradually reduce the nose-up angle. Correctly done, you will level off at the altitude you are supposed to be at. New pilots have a tendency to either overshoot or undershoot the desired altitude, ending up too high or too low. This is where the Vertical Speed Indicator (VSI) comes in handy in helping you judge when to begin the level off. A rule of thumb is to begin leveling off the plane when you are 10 percent of the VSI's reading below the desire altitude. Let's say we're climbing at 500 feet per minute and want to level off at 3000 feet. Ten percent of 500 feet per minute is 50. In this case, at 2950 feet we would start to level off, hopefully ending up at 3000 feet as the final altitude. If we were climbing at 1000 feet per minute, we would start to level off at 2900 feet.

When releasing backpressure on the control yoke, the input should be gradual. You don't want to force your passengers up against their seatbelts by shoving the control yoke forward too forcefully, giving them the feeling of weightlessness. Normally, a nice, steady release of the backpressure on the control yoke gives your passengers a pleasant transition from the climb to level flight. Once you are in level flight, you will need to reset your engine power, then retrim the elevators for level flight. If you retrim before resetting engine power you will probably end up retrimming again after the power change.

Descents are similar to climbs but have their own unique aspects. While some think descending is just a matter of shoving the nose down to lose altitude, getting down can show how much finesse pilots have when they fly. In this section we are going to discuss cruise descents, leaving landing descents for Chapter 6. Cruise descents differ from landing descents in a number of aspects, power settings and airspeeds being the most important. Let's assume we're flying along at 7000 feet AGL and we want to descend to 3000 feet. We're flying a single-engine airplane, with two passengers on board. In most cases, we will begin a descent by reducing engine power slightly; how much depends on the airplane we are flying, the acceptable airspeed ranges for the plane, and the weather conditions. If you are flying a plane that has a wide range of cruising speeds, and a descent at cruise power settings will not push the airspeed above acceptable limits, then it may be advisable to leave power settings alone and ease the nose down with forward control yoke pressure to begin the descent. But if the plane you are flying is more sensitive to airspeed ranges, and keeping cruise power on during the descent will drive the airspeed above recommended limits, then you will want to reduce power before beginning your letdown to avoid picking up too much airspeed.

Another consideration for descents is the weather. On a glass-smooth day, with very little turbulence, higher airspeeds within safe limits are acceptable. But on a day when you're being bounced around by turbulence, holding the airspeeds at lower values can help improve the ride for you and your passengers. Another consideration when flying in turbulence is the maneuvering speed of the plane, known as V_a. When flying in turbulence, it is always recommended that you fly the airplane at speeds at or below V_a. The maneuvering speed is the speed at or below which sudden or complete control movements will not cause structural damage to the airplane. At these airspeeds the plane will stall first, and keep the structural loads on the airplane within acceptable limits. However, if you are flying at speeds above V_a and encounter sufficient turbulence, it is possible to cause structural damage to the plane. As you can see, knowing the conditions you are flying in, and the plane you are flying, can help make the ride more comfortable and safe.

For our example let's assume we want to maintain V_a for the descent airspeed. We will reduce power, then lower the nose, using the elevator to control our airspeed and the power to control the rate of descent. This is an important concept to keep in mind as you fly. Airspeed is controlled by the elevator, and rate of descent, or climb, is controlled by the power setting. We will discuss this further in Chapter 6, but many pilots do not completely understand this concept, and you need to understand how to control airspeed and rate of descent or climb. If our initial airspeed in the descent is too high, we need to raise the nose of the plane slightly, which will reduce the airspeed; if the airspeed is too low, increase the airspeed by dropping the nose slightly. With airspeed established, you can then set your rate of descent using the VSI as an indicator. For most descents you should aim for approximately 500 feet per minute of descent. This gives you a descent at a controllable airspeed and does not become uncomfortable for your passengers. While you may need to increase the rate of descent due to various factors, a higher rate of descent causes your passenger's ears to pop more frequently, and if they have any head congestion it may make the situation even worse.

As with climbs, begin to level off about 10 percent of the descent rate before your target altitude. Ease back on the control yoke, slowly raising the nose. You can also begin to add power to help prevent the airspeed from falling off. Correctly executed, you can smoothly transition from the descent to level flight and hit your target altitude. Once you have leveled off and set your engine power, retrim the plane for level flight. So far we have discussed straight and level, climbs, and descents. Now let's move on to turns.

TURNS

We discussed turns in an airplane to some extent in Chapter 1 when we covered basic aerodynamics. In this section we are going to take a more detailed look at how to execute turns and how the plane reacts during them. As you will see, as the angle of bank changes, the way in which you fly the turn changes. We will also take a look at some common errors pilots make during turns.

As you already know, we use lift from the wings to actually turn an airplane in a properly executed, coordinated turn. We use ailerons to bank the airplane, which uses some of the lift available to turn the airplane. As we make the aileron input, we also use rudder in the same direction to overcome adverse yaw, which prevents the nose from yawing in the opposite direction of the turn. At the same time we input aileron and rudder, we increase backpressure on the yoke, increasing the angle of attack and generating more lift. This additional lift is needed to compensate for the lift we redirect to turning the airplane. Without this additional lift the plane would lose altitude.

All of this seems pretty straightforward. However, to make a smooth entry into and exit from a turn requires an understanding of the airplane you fly, a feel for the airplane and the feedback it gives you, and a certain amount of finesse on the controls. Let's begin with entry into the turn. Figure 2-6 shows the forward view looking from the pilot's seat over the nose of the plane in straight and level flight. The dot in the figure is an imaginary reference point on the windscreen that is directly in front of the pilot. As you enter the turn you want to maintain level flight and not climb or descend. The best visual reference you can have is the point on the horizon on the windscreen directly in front of you, which is represented by the dot. During turns, that point will remain at almost the same relation to the horizon above the control panel as it did in level flight.

Figure 2-7A illustrates when the plane is banked in a left turn. The dot shows where the horizon was during level flight. In this case the horizon has stayed on the dot, and the plane does not climb or descend during the turn. Figure 2-7B shows the horizon has moved below the dot, which means the nose of the plane has risen and the plane is climbing during entry into the turn. Figure 2-7C shows that the horizon has moved up in relation to the dot, meaning the nose of the plane has dropped and the plane is losing altitude. This visual reference can be verified by using the artificial horizon, altimeter, and VSI, which will also show when the plane is climbing or descending. A word of caution, though. Many student pilots become so focused on the instruments during turns that they stop looking outside the plane and stare intently at the instrument panel. When flying in VFR conditions, these instruments should verify what you are seeing by looking outside the plane, not act as your primary reference during turns.

Now that we've learned how to judge the proper horizon reference for pitch during a turn, let's talk about entry into the turn. Before you begin a turn, always clear the direction you are turning toward to make sure there are no other aircraft or other poten-

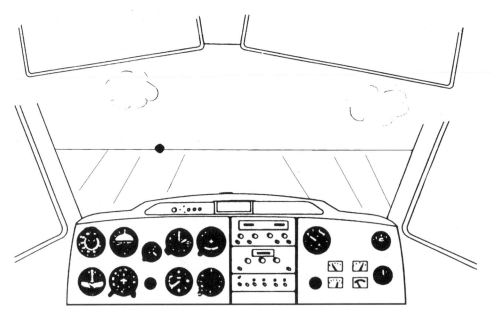

Fig. 2-6 *Forward view from pilot's seat (dot on horizon).*

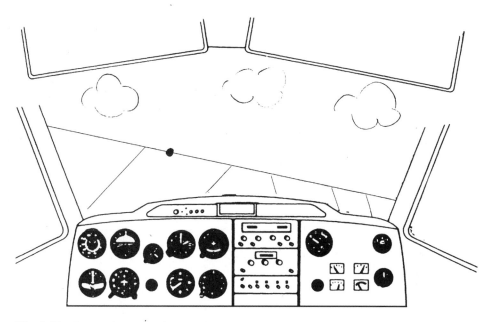

Fig. 2-7A *Forward view: Left turn (dot on horizon).*

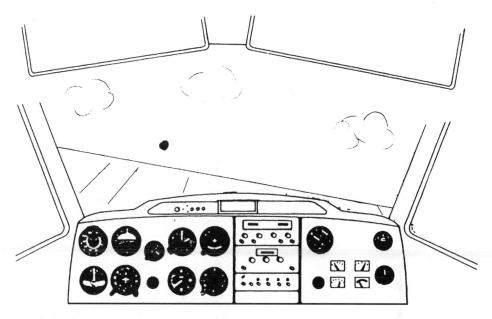

Fig. 2-7B *Forward view: Left turn (dot above horizon).*

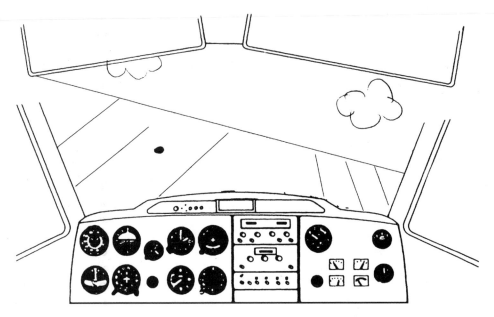

Fig. 2-7C *Forward view: left turn (dot below horizon).*

tial obstacles in that direction of flight. In a high-wing airplane this may require you to raise the wing slightly to clear the area. Once the turn area is cleared, you can look forward again and set your horizon reference point for attitude. Apply aileron in the direction of the turn, at the same time applying just enough rudder to prevent the nose from yawing and to keep the ball centered in the turn and bank indicator. You will also need to apply just enough up elevator to maintain your reference point on the horizon. It takes practice to become used to how much of each control should be used during a turn, and you will find that this changes not only from plane to plane, but also is affected by the airspeed you are flying at. We will discuss control use during slow flight further in Chapter 3, but at this point keep in mind that the slower an airplane is flying, the less effective the controls are. This means that you will need greater control movement to achieve the same results as compared to flying at higher airspeeds.

Once you are in the turn, you want to maintain a constant bank and altitude. By monitoring your external visual references, and verifying with the flight instruments, you can make minor adjustments to aileron, rudder, and elevator inputs. As you make any aileron inputs, you will also need to make corresponding rudder and elevator inputs. For instance, if you increase the angle of bank through greater aileron application, you will also need to apply more rudder. If you decrease the bank, you will need to reduce your rudder input. These inputs should be done at the same time, and the ball should stay centered.

To exit the turn you will roll the wings back to level flight, using rudder and aileron in a coordinated movement, just as when you entered the turn. As you reduce the bank, the lift vectors will again be used to keep the plane at altitude, and you will need to reduce backpressure on the control yoke. When entering turns it is very common for students to either not increase backpressure on the control yoke, and the airplane begins to lose altitude, or to overcompensate and add too much backpressure, which causes the plane to gain altitude during entry into the turn. When exiting the turn it is very common for students to leave the elevator input in, which causes the plane to gain altitude during the rollout to straight and level. Another common problem for students is adding rudder after aileron is rolled in, which allows the nose of the plane to yaw visibly in the opposite direction of the turn. Not only does this make the entry into the turn less smooth, but it also changes the reference point's position as the nose of the airplane moves and makes it more difficult to maintain the reference point with the horizon. This often results in the plane climbing or descending during entry into the turn.

There's also another important factor you need to take into consideration during turns. In a shallow banked turn, from 0 to 20 degrees bank, the dihedral of the wings will attempt to restore the plane to wings-level flight, so you will need to keep ailerons input throughout the turn. In a medium banked turn, 20 to 45 degrees, the plane will have a tendency to remain in the bank, and you will be able to neutralize the ailerons to some extent once the bank has been established. In a steep banked turn, greater than 45 degrees, the plane will have a tendency to roll steeper, and you may need to feed in some opposite aileron to counteract this tendency. Remember, as you make these varying aileron inputs you will also need to adjust your rudder inputs to maintain coordinated flight.

Steep Turns

In steep turns you will find that it requires more effort to maintain altitude. Because so much of the lift is vectored to turn the aircraft, the increase in the angle of attack must be greater as compared to shallower banks. You will find in some planes that not only do the elevator inputs become fairly heavy, but also the plane may not be able to maintain altitude unless additional power is added during entry to the steep banked turn. By adding power you will have more available lift, and it will be easier to maintain altitude. With some planes, no matter how much power you add you will find they have a tendency to lose altitude once the bank exceeds a certain angle. You do not normally trim the airplane for turns, since they are normally done over a short time frame and elevator pressures are held for only a short period of time.

A common error that students tend to make during steep turns is to become uncoordinated as the bank becomes steeper. When in a steep turn, pilots will hold rudder opposite the direction of the turn, which tends to yaw the nose of the plane up slightly. For example, in a 60-degree left bank turn, some pilots hold right rudder to help hold the nose of the airplane above the horizon. This is often a response to the heavy elevator pressure they are holding and is often an unconscious act on their part, but this places them in an uncoordinated turn.

In Chapter 3 we will discuss stall speeds in detail, but you should be aware of the fact that as the angle of bank of a plane increases, the stall speed of the plane goes up. It is quite possible to put a plane into a high-speed stall in a very steep bank. If the pilot has the plane in an uncoordinated steep turn when it stalls, it is quite possible the plane will enter a spin. I enjoy doing spins, and I teach them on a regular basis to students. But an unplanned spin from a steep turn can make for an exciting day for you and your passengers. If you are uncomfortable with the elevator pressures you must hold to maintain altitude during a steep banked turn, reduce the angle of bank to make things more controllable.

As you roll out of a steep turn you will need to remember to reduce the engine's power setting, and to reduce the backpressure on the control yoke. There is a tendency to gain a great deal of altitude while exiting a steep turn as a result of the additional power and increased elevator backpressure.

AIRSPEED AND CLIMB OR DESCENT CONTROL

Earlier in the chapter we touched on the fact that you use elevators, or pitch angle, to control your airspeed and power to control your rate of climb or descent. At this point we are going to take a more in-depth look at that concept. It is a basic one which is misunderstood by many pilots and needs to be clearly recognized in order to master many aspects of flying an airplane well.

You may be asking yourself, "How do the elevators control airspeed? They control the pitch of the plane, which is how I climb or descend." That is partially true, elevators do control the pitch of the plane, which directly controls the speed the plane is flying at. What happens when you are driving along in your car and start up a steep hill? You begin to slow down. As you crest the hill and start down the opposite side, your car begins to increase in speed. This is much the same way pitch controls airspeed as you fly. When you are climbing, you are fighting the force of gravity, which acts as a brake on the plane's speed. If you are going too slowly, you reduce the pitch angle, which results in an increase in airspeed. When an airspeed is too high, increasing the pitch angle slows the airplane. This is true during both climbs and descents. When we discuss landings, this becomes especially important. To fly a good, consistent pattern you need to maintain constant airspeeds and rates of descent. Holding a constant pitch angle will help you fly at a fixed airspeed and make it easier for you to hit your landing target consistently. So, if you are too slow, reduce your pitch angle. If you are too fast, increase the pitch angle.

Now that we know how to control airspeed, let's discuss how to control your rate of descent or rate of climb. The other control we have at our disposal is the throttle, which controls engine power. In most climb situations we will be using a specific airspeed, such as V_x or V_y, as the target we are using. The rate of climb will be dependent on the airspeed, how the plane is loaded, and atmospheric conditions. Because of this the rate of climb will vary quite a bit from flight to flight, or even within the same flight, depending on conditions. I have spent quite a bit of time flying skydivers to altitude, only to watch them jump out of the airplane, land, and stand in line to do it again. The limiting factor was how quickly I could get the plane back down and then get them back to altitude again. After takeoff I would establish V_y, the best rate of climb, to help avoid overheating the engine. Since we were taking off less than 1000 feet above sea level, density altitude not withstanding, the initial climb rate was normally good. But as the plane continued to climb, the vertical speed indicator would show that the rate of climb continued to drop. On many days as the plane passed 7000 feet on the way to 10,500 feet AGL, the rate of climb would be only two or three hundred feet. The last thousand feet to the jump altitude would often take several minutes, with me coaxing it all the way. Unless you happen to be flying an F-14, the challenge is not to exceed the climb rate you are looking for, but to maintain adequate climb rates. Hot summer days make the problem even more interesting. Use the aircraft manufacturer's recommended airspeeds and power settings and you should get the best climb rates available.

While we may often wish we had to control climb rates, we should always work to control the rate of descent we use. Airspeed is controlled with the elevator, and we will control the rate of descent with power. In Chapter 9, which covers basic instrument flying, we will learn that instrument flight often demands that we use a specific rate of descent to maintain the correct glideslope. When flying in VFR conditions this may not be a requirement, but passenger comfort will be higher if you do not descend too rapidly. Additionally, controlling the rate of descent will help avoid an unplanned increase in airspeed. Conversely, descending too slowly may put you too close to other aircraft if you are working with air traffic control and they are asking you to descend for separation from other airplanes. Descending too quickly with a low power setting can also cool your engine very rapidly. We know that climbing for extended periods at too great an angle can overheat the engine and reduce engine life. Cooling the engine quickly can have a similar effect, the rapid contraction of metal parts putting unnecessary stress on the components.

Unless there is a reason to descend more rapidly or slowly, I have found that descents of about 500 feet per minute seem to work well for a number of reasons. First, you do not need to reduce power a great deal to keep the rate of descent constant, and it is relatively easy to control the airspeed with pitch. Second, the rate of descent is not so great that it causes the passengers discomfort. Passengers who do not fly frequently may not be accustomed to clearing their ears, and having to do so frequently as the result of a rapid descent may be unpleasant for them. Finally, with moderate power settings you can still avoid cooling the engine too rapidly, and put less wear and tear on it.

In a descent you will first want to lower the nose and reduce power slightly. The airspeed will be controlled with the pitch. If you are too fast, reduce the nose-down angle; if you are slow, increase the pitch angle. To hit the target rate of descent you are looking for, increase or decrease power accordingly. If the rate of descent is high, increase the power setting. Be aware that any change in power setting will also require a change in pitch angle to hold the correct airspeed. To increase the rate of descent, you will reduce power and change the pitch to hold a constant airspeed. As you execute a descent you may also want to reduce your airspeed to V_a or below, depending on the turbulence. Even if you start out in smooth air it is quite possible that as you get to lower altitudes that the air could become more turbulent. How many summer afternoons have you been at 6,000 feet AGL in smooth, calm air, only to get bounced around the cockpit as you descend to land. Be aware of the environment and how a climb or descent can require that you adapt to meet new flying conditions.

This sounds like a great deal of making power and pitch changes to hold a constant rate of descent, but with a little practice you will find that it becomes very easy to maintain both a constant airspeed and the rate of descent. The key is knowing the airplane that you are flying and the appropriate pitch angles and power settings. Anticipating what you will need to do reduces the workload and the need to search for the right configuration.

CONCLUSION

This has been a chapter that will be the foundation of many of the concepts we cover throughout the rest of the book. Being able to hold straight and level flight and knowing how to correctly turn the plane and execute climbs or descents are the basics of flying. As you learn to fly, or work to improve your flying skills, you will find that if you are having problems with some of the more advanced skills, it often stems from poor technique in the flying skills we have covered in this chapter.

Take the time to critique yourself as you fly; notice if your turns are coordinated and smooth. Do your climbs and descents end up at the altitude you are aiming for? Does the airplane wander from the correct heading or altitude? When you are completing a turn, do you roll out on the correct heading? If you find that you may not be crisp on hitting the mark when executing these maneuvers, it might be time to focus on the basics we discussed in this chapter. It may even be worthwhile to take along your favorite flight instructor for a little brushup on the basics of flying. If you work to perfect the topics we covered in this chapter, you will find that it carries through to all of the flying you do. And after all, the point of flying is to be the best you can be at it.

3
Slow Flight and Stalls

NOW WE ARE GETTING INTO FLIGHT MANEUVERS THAT I REALLY LIKE flying. I feel that the more comfortable that pilots feel flying slow flight and stalls, the better they will be at avoiding unplanned stall situations. Whether I am teaching aerobatics or taking a pilot up in a Cessna 152, I am a big believer in giving slow flight and stall training. Slow flight teaches a pilot how an airplane behaves at slower airspeeds, how to use the controls, the difference in control authority, and other noticeable changes between slow flight and flight at normal cruise speeds.

In my opinion, stalls are even more fun to fly. They take the airplane into the flight regime just past slow flight and build on the skills you learn as you practice slow flight. The true desire behind this chapter is not necessarily to convince you to like doing stalls, although I hope that is one outcome, but to help you learn to recognize the onset of stalls and to be able to avoid them in critical situations like takeoff or landing. The NTSB has far too many accident reports that end in a stall/spin situation during takeoff or landing. If the pilots of those flights had been more familiar with slow flight and stalls, it is quite possible that at least some of those accidents might have been avoided.

As you read this chapter, keep in mind that if you are not familiar with the maneuvers, or feel at all uncertain about flying them, take a qualified flight instructor up with you to practice them, or any other topic discussed in this book. Let's begin by looking at the principles of slow flight.

PRINCIPLES OF SLOW FLIGHT

Slow flight can be defined as maneuvering the airplane at airspeeds close to the stall speed of the plane. I have heard the definition of slow flight includes airspeed ranges for a given aircraft, but for the private pilot flight test. The examiner is looking for speeds 5 knots above stall speed during several maneuvers centered on slow flight. In this section we are going to cover how control effectiveness changes during slow flight, some left turning tendencies that can be present during slow flight, and how to enter into and recover from slow flight. We will also discuss flying several maneuvers during slow flight, such as turns and straight and level flight.

Control Effectiveness

During flight at cruise airspeeds the controls of most general-aviation aircraft have a solid, authoritative feel to them. When you roll in aileron, or use the elevator, the plane reacts in a crisp, responsive manner. As you slow a plane down, though, the airflow over the controls is reduced. As a result, the controls become less effective and acquire a "mushy" feeling. You will find that they not only feel softer due to the lower airspeed, but also the plane is much less responsive to control inputs and it takes a greater amount of control input to get the plane to execute the maneuver you are attempting to fly. All of this is tied to the lower airspeed you are flying at.

When you are in slow flight mode, controlling the airplane becomes a balance between feeling the controls, noting their effectiveness, not overcontrolling due to their loss of authority, and judging how much control input is necessary to accomplish the maneuver you are flying. It is quite common for pilots to overcompensate control inputs as the controls become less effective. In these cases the plane can wallow, or oscillate from these pilot-induced inputs, never quite holding the correct attitude. Instead, the plane moves through the desired attitude in one direction, then back through it in the other, either rolling, yawing, or pitching as the pilot overcompensates.

When you are in slow flight mode, make the control inputs relatively gently, keeping in mind how the feedback on the controls is and how the plane reacts to the inputs. Depending on your familiarity with the aircraft, you may need to build a "feel" for control inputs at different airspeeds, from cruise down to just above stall speed. This will allow you to be able to anticipate how the airplane is going to react when you are flying at different speeds. Go out and practice flying at slow airspeeds until you can make smooth, coordinate inputs on the controls. As you become more practiced at slow flight you will find you can hold the airplane at just the desired attitude with smooth, coordinated control inputs.

As you will see later in the chapter, overcontrolling the plane at slow speeds can result in a stall/spin situation, while using the controls correctly can help you avoid an accidental stall/spin during takeoffs and landings. The other benefit of learning the feel of the controls at slow airspeeds is that, based on how the controls feel, you can begin to tell when the plane is slowing down without looking at the airspeed indicator. This can be very helpful in situations where you are distracted while flying and may not immediately notice that you have raised the nose and that airspeed is bleeding off. In this case, noticing that the controls are becoming mushy can alert you to the fact that you are losing airspeed.

Left-Turning Tendencies

When most standard-configuration general-aviation planes are in slow flight mode, they have a tendency to yaw to the left. For the purposes of this discussion we are concerned with single-engine aircraft that have the propeller turning in a clockwise rotation as seen from the pilot seat. If you happen to be flying a pusher aircraft, or one that has the engine turning counterclockwise, the opposite effect may take place, resulting in the plane yawing to the right.

The left-turning tendency is actually due to four factors: gyroscopic precession, P-factor, slipstream, and torque. Each of these factors can add to the tendency of the plane you are flying to yaw to the left when you are flying along at a slow airspeed, such as during takeoff or landing. Knowing how to use the controls to compensate for these tendencies can help you anticipate the control inputs you will need and maintain coordinated control inputs.

You may be asking yourself why compensating for left-turning tendencies is important, but there are several reasons you want to make up for them correctly. First, if you are flying along in a slow flight mode and not compensating for the left-turning tendencies, if the plane stalls it may enter a spin due to the uncoordinated use of the controls. If the plane does not enter a spin, it may have a tendency to drop a wing as the plane stalls, which an unwary pilot can aggravate if the wrong control corrections are applied. Even something as seemingly simple as takeoff can be complicated by the four left-turning tendencies. I know of a flight instructor who was giving a student soft-field takeoff instruction that ended up having the airplane veer sharply to the left as the nose of the plane was raised during takeoff. As a result the plane left the runway, dug into some very soft ground, and flipped over on its back. Using sufficient right rudder during rotation might have helped prevent the plane from yawing to the left and avoided a potentially dangerous situation. Let's take a look at each one of the left-turning factors, beginning with torque.

Torque

Figure 3-1A shows how a plane has a tendency to roll left as a result of the torque created by the engine. When the propeller of the plane you are flying rotates, it creates an equal and opposite reaction in the opposite direction. In most single-engine general-aviation aircraft, the propeller turns clockwise as you are sitting in the pilot's seat. As a result the airplane has a tendency to rotate in the opposite direction, in this case a rolling tendency to the left. The tendency becomes more pronounced the slower the plane is flying at, and the higher horsepower the engine is generating. As you can see, takeoff situations present a prime opportunity for the left-rolling tendency to be the greatest. I have read that some World War II fighter aircraft, the P-51 in particular, must have the tailwheel locked in place before they begin the takeoff roll, or they cannot keep the plane from turning to the

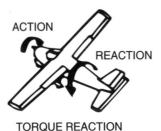

ACTION

REACTION

TORQUE REACTION

Fig. 3-1A *Torque reaction.*

left due to engine torque. They are generating so much horsepower that flight controls at low airspeeds cannot prevent the turn.

Those of us flying a Cessna or a Piper are not faced with the same degree of challenge in keeping the nose pointed straight down the runway as those lucky few flying such high-horsepower airplanes. But we must still understand what is taking place and correct for it. Even something as subtle as tire wear can be affected by engine torque. I fly a Pitts S2-B, which generates 260 h.p. The left tire has a tendency to wear down much faster than the right tire due to the extra pressure put on it by the torque of the engine during takeoff. Some highly modified racing aircraft employ contra-rotating propellers to help offset the left-turning tendency caused by torque. The racing P-51 Miss Ashley has two three-bladed propellers to reduce the effect of torque on the plane. Other aircraft over the years have also adopted this technique, but the cost and complexity of the solution reserve it for very special situations.

It may take a combination of right rudder and right aileron to compensate for the left-rolling tendency the torque of the engine generates. The greater the horsepower the engine is capable of producing, the stronger the tendency to yaw to the left will be. If you happen to fly an airplane that has an engine turning in the opposite direction, such as a Russian Sukoi 31, with its big radial engine, or even some of the Rotax powered airplanes, you will find that the plane has a tendency to roll to the right. As a result, you will need to compensate with left rudder and aileron inputs to maintain coordinated, level flight.

Slipstream

The next left-turning tendency we will cover is slipstream. This force is generated by the airflow created by the propeller. Figure 3-1B illustrates how the air flowing back from the propeller as it turns has a corkscrewing motion around the fuselage of the airplane as it moves toward the rear of the plane. This rotating air strikes the vertical stabilizer on the left side of the plane, pushing it to the right and causing the nose of the plane to yaw to the left. Like torque, the slower the airplane is and the more horsepower it is generating, the stronger this left-yawing tendency will be. It will take a certain amount of right rudder to compensate for this yaw to the left, which is most pronounced during slow flight and high engine-power-setting situations.

As the plane you are flying gains additional airspeed, the slipstream lengthens, which reduces the pressure it places on the vertical stabilizer at cruise airspeeds. But as you slow the plane, the slipstream shortens and is able to impact the vertical stabilizer with more force. Again, this is why the plane has a more pronounced tendency to yaw to the left during slow flight situations such as takeoff.

Gyroscopic Effect

Gyroscopic effect is the third factor that we will discuss that causes an airplane to yaw to the left. Figure 3-1C shows the effect that placing a force against a gyroscope can have. Essentially, when you place a force perpendicular to the plane of rotation of a gyroscope, the resulting motion occurs 90 degrees in the direction of rotation from the point the force was placed. The figure illustrates this point, showing that the gyroscope actually moves in a direction 90 degrees from where the force was input.

The same principles of physics hold true for the propeller of the plane you fly. In reality, the propeller is actually a large gyroscope, which has a great deal of mass and spins very rapidly. When you make pitch changes to the airplane, it is the same as inputting a force against a point on the gyroscope. In planes equipped with tailwheels, the plane will want to yaw to the left as you raise the tail of the plane during the take-off run. The Pitts Special I fly has a very definite yaw to the left as I raise the tail and requires that right rudder be used to keep the nose pointed straight down the runway. The strength of the yawing effect will depend on the weight of the propeller and the speed at which it is turning, as will the amount of rudder needed to compensate for the yaw it generates. You should be aware that any pitch change results in the same gyroscopic effect, not just during takeoff or slow flight. But these are the times when it can be the most critical due to the need to maintain coordinated use of controls, which we have discussed in depth.

SLIPSTREAM

Fig. 3-1B *Slipstream.*

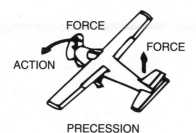

PRECESSION

Fig. 3-1C *Precession.*

Chapter Three

P-Factor

P-factor is the last left-turning tendency we will discuss. When the blades of the propeller turn, they are actually generating lift just as the wings of the plane you fly do. As the angle of attack increases, so does the lift that they generate. The same holds true for the blades of the propeller. Figure 3-1D shows the nose of a plane in the climb attitude. As you can see, during the climb the descending propeller blade, which is on the right side of the aircraft, is at a higher angle of attack than the ascending blade. Due to this higher angle of attack, it generates more thrust than the blade on the left side of the aircraft. This difference in thrust between the left and right sides of the propeller results in a tendency for the plane to yaw to the left.

Like many of the other left-turning factors, the higher the horsepower the engine is generating, the greater this tendency will be. It is most pronounced when the plane is creating high horsepower at slow airspeeds, such as in slow flight mode or during takeoff. For this reason you will need to input right rudder to counteract this left-yawing tendency.

What many pilots are not taught is that when the nose of the airplane is down, such as in a descent, the plane will have a tendency to yaw right for the same reasons. The propeller blade on the left side is now generating more lift than the one on the right, and the nose of the plane will yaw to the right. This will require that you input left rudder to keep the nose of the plane from yawing. However, since most descents are made at cruise or reduced power settings on the engine, you will find that the effect is also reduced and not as pronounced as during takeoff or slow flight.

Now that we have covered the four left-turning tendencies, you may be wondering why the plane you are flying does not yaw to the left all the time. The engine is still generating torque and slipstream at cruise power settings, yet you do not need to continuously use right rudder to overcome these factors. To compensate for the left-turning tendencies during cruise flight, the engineers who design airplanes have incorporated several design features into how the airplane is built. These design features compensate for torque and other yawing tendencies. For instance, the vertical stabilizer on many airplanes is offset to the left of parallel with the longitudinal axis of the plane by several degrees. This offset compensates for some of the left-yawing tendency a plane may have. The engine of the plane you fly is also offset to the right by several degrees, again to compensate for any left-turning tendency at cruise flight configurations. This is accomplished in the construction of the engine mount, which angles the engine slightly to one

ASCENDING DESCENDING NOSE HIGH
LEFT BLADE RIGHT BLADE ATTITUDE

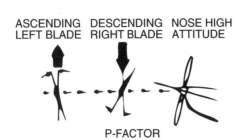

P-FACTOR

Fig. 3-1D *P-factor.*

side. Finally, the engineers can design a twist along the length of the wings, known as washout, to help reduce any rolling tendencies due to torque. Washout creates different angles of attack at points along the wing as a result of this twist. This twist may differ between the wings to compensate for the left-rolling tendency caused by torque. Washout is also designed to help with stall characteristics, allowing inboard portions of the wing to stall before the outboard portions.

At this point we have a good understanding of changes in control authority during slow fight and the tendency of the plane to yaw to the left in that flight regime. Let's take a more in-depth look at slow flight.

Slow Flight

Slow flight is an interesting flight regime. You are slowing the airplane to just a few knots above stall speed, the controls become mushy, you can begin to feel the onset of buffet due to the turbulent airflow over the wings, and it takes a certain amount of finesse to smoothly fly the airplane through maneuvers. I like training students to feel comfortable with slow flight because it makes them so much more aware of what the plane is doing in this portion of the flight envelope. Having a comfort factor with slow flight is extremely useful during landings; you can feel the airplane, anticipate its reactions, and as a result fly more safely during the landing phases.

In this section we are going to discuss slow flight entry and recovery from slow flight, including common errors pilots make during slow flight. When you are done with this section, you should have a solid understanding of how to reduce the airspeed of the plane down to slow flight speeds while maintaining a constant altitude, how to maneuver the plane through turns in the slow flight mode, and how to recover from slow flight. As you will see, the only thing you need to do to become proficient with slow flight is practice.

Slow Flight Entry

Before we get into a discussion of slow flight entry, a word of caution. Any time you are practicing slow flight, you should be sure you have sufficient altitude to recover from any unplanned stalls or spins. I have read accident reports from the NTSB where pilots were flying at low altitudes in slow flight mode and ended up accidentally spinning the airplane without sufficient altitude to recover. When I practice slow flight with students, we usually fly at least 3000 feet AGL, so please keep safety in mind as you practice this or any other maneuver.

Slow flight entry is straightforward. For the purposes of our discussion in this section, we will assume you are entering slow flight from straight and level attitudes at cruise airspeeds. The idea of slowing to the slow flight regime is to maintain heading and altitude while slowing the airplane to about five knots above stall speed. The plane you are flying may differ in the procedure to slow it down, so be sure to check the manufacturer's recommendations before you begin practicing.

Essentially, if you start in cruise configuration, begin slowing the plane by reducing power slowly. In airplanes like a Cessna 152 or a Piper Cherokee, I like to start by reducing the engine's power to about 1700 r.p.m. on the throttle setting, then adjust as necessary to maintain altitude when the desired airspeed has been achieved. As the plane begins to slow after the initial power reduction, ease back on the control yoke to slow the plane and maintain a constant altitude. How much and how quickly you will need to input back-pressure on the control yoke will depend on the plane and how much power has been reduced. As the plane slows and the pitch of the nose is raised, you will need to input right rudder to keep the ball centered as the left-turning tendencies begin to come into play. Many students forget to make the rudder correction, and the nose of the plane begins to drift slowly to the left. As a result, the plane no longer holds the heading they started out on as they entered slow flight. It is also common for students to lose altitude while the plane slows. Be sure to increase the pitch of the plane enough to help slow it down and to maintain altitude as it slows.

As you begin to reach the target airspeed, it might be necessary to adjust the power setting up or down to allow you to maintain altitude. Remember, if the airspeed is too high, you will slow the plane by increasing its pitch angle. If the speed is too low, reduce pitch angle. Each of these inputs will result in a change to your vertical speed as well, so you will probably need to adjust your power setting in conjunction with any pitch change.

With a little practice you will be able to anticipate the correct power and pitch inputs, and slow the plane to the correct airspeed while maintaining altitude very consistently. If you want to use flaps as part of the slow flight mode, add them once the plane has slowed to the white arc of the airspeed, then apply flaps in increments. I have found that applying flaps one notch at a time gives you a smoother transition to slow flight. You can use any flap settings, from no flaps to full flaps during slow fight practice. I recommend that you become proficient with each flap setting, building a "feel" for the plane at each of these configurations.

When you are adding flaps, be prepared for the pitch change that will accompany their deployment. For most single-engine airplanes, as you add flaps they will pitch nose down; how much depends on the design of the plane. Each airplane will react differently, so learn the plane you are flying and become familiar with its flight characteristics.

At this point we have slowed the airplane, using pitch to control the airspeed, and have applied the flaps we want to use for this practice session. We set power at an initial setting, 1700 r.p.m. for this example, now what do we do next? In Chapter 2 we covered using the pitch angle to control airspeed and using power setting to control rate of climb or descent. Now we get to put that information to practical use. If we have reached our target airspeed, approximately 5 knots above the stall speed, but we are unable to maintain the proper altitude, we need to adjust the power setting. If the plane is climbing, you should reduce power slowly. Now here is where we have to work the controls in concert to keep things in balance. As a result of the change in power setting, the airspeed will begin to bleed off. To offset that tendency, you will need to reduce backpressure on the control yoke and reduce the pitch of the plane. This will allow you to maintain a constant airspeed while the plane adjusts to the new power setting. Until you become familiar with a plane's flight characteristics, you may need to go through several iterations of power reduction/pitch change to get the plane to hold the proper altitude and airspeed.

The same holds true if the plane is losing altitude after the initial slow-flight power setting is established. If the plane is descending and you have the proper airspeed dialed in, you will need to increase your power setting slowly. Just as with the power reduction, you will also need to make a corresponding pitch change to maintain a constant airspeed. With an increase in power settings, you will need to increase the plane's pitch to maintain the desired airspeed. By making power changes slowly, you will avoid overcompensating with too much or too little power.

I find that holding airspeed is easier for most students than holding altitude. The problems begin when they fail to use power and pitch together when making power or airspeed changes. Just as changing power setting necessitates that you make pitch changes, when you change the pitch of the plane to increase or decrease the airspeed, you will also need to adjust the power to hold altitude. Generally, if you reduce the pitch to increase airspeed, you will need to decrease power to avoid gaining altitude. If you increase pitch to slow the plane, you will likely need to increase power to avoid an altitude loss. Remember, use the controls together and anticipate what you will need to do to maintain airspeed and altitude.

Once you are set up in slow flight mode, you will probably make almost continuous changes in control inputs to hold your airspeed. Power settings should remain fairly constant once you have things close, with only minor power-setting changes necessary in some cases, which we will discuss in the next few paragraphs. You will find that minor pitch changes will be necessary to maintain airspeed, though. These are small changes and will have minimal impact on the need for power-setting changes.

Finally, we have achieved slow flight mode and are maintaining a constant airspeed and altitude. What do we do now? The one thing you should notice immediately is the softness of the controls. Ailerons are mushy, with relatively large inputs needed to hold the wings straight and level as compared to cruise airspeed. The elevators are also less authoritative, and you will find that you will need a fairly large amount of right rudder to overcome the left-turning tendencies we discussed in the last section. The noise of the plane has also changed. The engine is producing less power, and the sound of airflow over the surface of the plane is also quieter at slower airspeeds. Get the feel for all of these sensations because you can detect changes in airspeed from these sensory inputs without ever looking at the airspeed indicator. As we already noted, sensing these types of changes can help you realize you are slowing down if you are distracted from the instruments by some activity, helping you avoid an accidental stall/spin situation.

We can also perform a number of skill-building exercises during slow flight; altitude and heading control are important, as are maintaining coordinated use of the controls and keeping the wings level. However, one of the most useful exercises is making turns while in slow flight mode. Here is where we have to start thinking about stall speeds and what effect banking the airplane can have on them. Figure 3-2 depicts the load factor placed on the plane in relation to the angle of bank of the plane. As you can see, as the bank of the plane increases, the load factor also increases. This increase results in the plane needing to generate more lift to compensate for the additional load placed on the wings. This increase in load as a result of a bank results in an increase in the stall speed of the airplane,

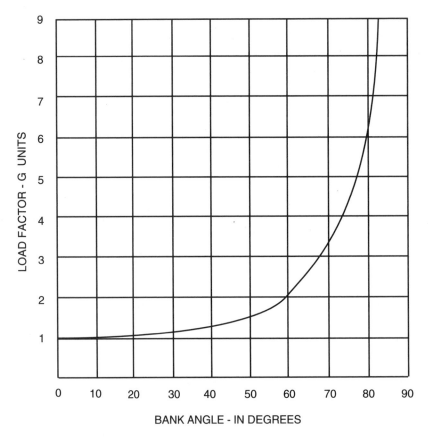

Fig. 3-2 *Load factor chart.*

something we need to understand especially well when we are in the slow flight mode. If we are flying just five knots above stall speed, and we attempt to make a turn at a 45-degree bank, we are quite likely to end up in a stall due to the increase in stall speed as a result of the bank. The slower we fly, the more likely we are to stall during a steep turn as a result of this relationship between load factor increase and a corresponding increase in stall speed. To finish the thought, this puts us in a particularly sensitive situation during takeoff and landing, when our airspeed is lowest and we are maneuvering around the pattern. All too often a pilot who is overshooting the turn to final may try to force the plane around by banking very steeply, causing a sharp rise in stall speed at slow airspeed and close to the ground. This is a common theme in many NTSB accident reports, so keep in mind that your stall speed increases as you increase the bank of the plane. This alone is a good reason to never exceed the 30-degree bank recommendation in the pattern.

Let's discuss turns to the left and right in the slow flight mode. They are flown just as we would when making a turn in cruise flight, but we must realize the controls will be less effective, and you will need to keep your banks shallow to avoid a stall. You will probably need to keep right rudder inputs applied, even in turns to the left, to keep the ball centered. In slow flight, you can feel a certain amount of control buffet, which also adds to the sensations. The buffet may increase during the turn as a result of airflow disruptions and the increased load as a result of the turn. You will also need to monitor your airspeed and altitude during turns. The plane has a tendency to lose altitude and airspeed during a turn, and it may be necessary to add power and adjust the pitch to keep your airspeed and altitude locked in. The amount of the power and pitch change will depend on the plane and how steep the bank is, but you should practice at different bank angles to become familiar with these changing settings. As you roll out of the turn you will need to reset power and pitch to avoid gaining altitude or airspeed as the load on the wings is reduced. A final note about turns in slow flight. In turns to the right you will find that a large amount of right rudder is necessary to keep the ball centered. Not only are you trying to overcome the left-turning tendencies of the plane with right rudder, but you are also overcoming adverse yaw, which was discussed in Chapter 1. The fact that the rudder is less effective at slow airspeeds also results in a need for even more input to keep the ball centered.

Once you become comfortable with turns in slow flight, I highly recommend practicing banking the airplane to the point of causing it to stall, to help you learn how the plane feels and reacts in these types of situations. As with any maneuver, you should have a qualified flight instructor along until you become familiar with the maneuvers. The instructor can help you learn and critique your performance and help prevent you from getting into trouble.

Our next area of interest is recovery from slow flight mode. The idea is to recover back to cruise power settings without gaining or losing altitude and maintaining the correct heading and attitude. The first step you will need to execute is to add engine power slowly, back to cruise power settings. As you add power you will need to reduce backpressure on the control yoke to avoid gaining altitude. If flaps were extended during entry into slow flight, you will want to remove them in increments. The reason for the incremental approach is to avoid a sudden loss of lift, with the associated risk of stall or nose-down pitch movement from the plane. After each notch of flaps is retracted, give the plane a few seconds to gain additional airspeed before retracting the next notch. If you have extended landing gear, you will also want to retract them during this time. One thing you should be cautioned about is exceeding either the maximum flap extension speed, V_f, or the maximum speed the landing gear can be down or retracted at. If these airspeeds are exceeded while the flaps or landing gear are extended, you may damage the structure these components are attached to, or the components themselves. As the flaps are fully retracted and the gear, if any, is retracted, you will find that the plane accelerates to cruise airspeeds fairly quickly and you will need to make sure you ease forward on the yoke to allow the airspeed to increase, and not use that power to climb. Here is a last note: When you enter slow flight mode, you should trim the elevator to a neutral pressure to avoid fatigue and allow for lighter control inputs. As you exit slow flight mode, you will also need to reset your trim settings to reduce the elevator input pressures there.

At this point you should have a good understanding of how to enter slow flight, what to be concerned about as you maneuver in slow flight, and how to exit slow flight. Let's take a look at some of the common errors pilots tend to make when flying slowly. First, they commonly forget to use enough right rudder to keep the ball centered in the glass tube. A large number of pilots I have flown with have become fairly complacent about using the rudder pedals correctly to overcome the left-turning tendency found during slow flight. Remember, keep the ball centered. If the ball is centered, the plane cannot enter a spin, and you can recover from a stall with less altitude loss than you can a spin. You will also find that if you do not keep the ball centered, the plane has a tendency to constantly turn to the left, making it difficult to maintain a given heading.

Another common error is erratic use of the controls during slow flight. Due to the lack of control authority, pilots tend to overcompensate, causing the plane to wallow around instead of holding a constant attitude. Make your control inputs with finesse anytime you are flying, but especially when you are flying at slow airspeeds. As you will see later in this chapter, misuse of the controls and overcontrolling can aggravate a stall situation. Learn to feel the airplane's feedback through the controls; this can tell you a great deal about how close the plane is to a stall.

Pilots who do not practice slow flight often enough also have a tendency to overcontrol with the throttle. Once you get the correct throttle setting, you should only need to make very minor changes in the power setting to control altitude. Too many pilots make their changes to the power settings larger than they should, resulting in a patterning of adding power, then pulling power back, then adding it again. Besides being tougher on the engine, it makes it difficult to let the plane reach a balance of the four forces we discussed in Chapter 1: lift, gravity, thrust, and drag. If these forces do not reach a balanced state, it becomes more difficult to control the airplane smoothly. Now that we have a basic understanding of slow flight, let's move on to the next maneuver we are interested in—stalls.

STALLS

In my opinion, stalls are one of the least understood maneuvers executed, or not executed, by general-aviation pilots. The curriculum for the private pilot tends to minimize actually performing a full stall, and an unfortunate number of flight instructors seem to be happy to teach only "imminent stalls," or slowing the plane to the point that the plane is about to stall, but never actually letting the stall develop. I am surprised by the number of private pilots who have told me they have rarely, if ever, done a full stall, because the primary flight instructor seemed to be very concerned with executing one.

If done at a safe altitude, with the proper training, stalls are as safe as any maneuver a pilot will execute. I stress the proper training point because this is very important. As we covered in the slow flight section, if you are not taught the proper use of controls, a stall can develop into a spin. Spins are fun to execute, if you are planning on doing them and you are in the right airplane to execute them. But if you get into one without the proper training, it can make for an exciting day. In this section we are going to look at the various types of stalls you should be aware of, including normal and accelerated stalls and approach and departure stalls. We will also discuss stall avoidance issues and how to minimize altitude loss during stall recovery. Finally, we will cover the common errors students tend to make during stall entry and recovery. When you have completed this section you should have a sound understanding of what a stall is, how it develops, and how to recover from it. As always, before practicing stalls on your own, take a qualified flight instructor up who can help you become proficient with these maneuvers. It will make you a safe pilot and save you time in learning the maneuvers.

Stall Definition

In this section we will define what a stall is and what causes one. As part of this discussion we will also review what the angle of attack is, in addition to the critical angle of attack. While these were discussed to some degree in Chapter 1, both of these pieces of information are critical to understanding how and why an airplane stalls. Let's begin with the angle of attack.

Figure 3-3 shows an airfoil as it moves through the air. The flow of air that moves directly toward the airfoil as it travels through the air is known as the relative wind. An arrow representing relative wind can be seen in the illustration. This forms an angle with the cord line of the wing; this angle is known as the *angle of attack*. As you know, as the angle of attack increases, so does the lift the wing produces, up to a point. Figure 3-4 shows several illustrations of a wing and the airflow over it at

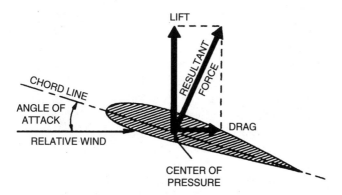

Fig. 3-3 *Relative wind/angle of attack.*

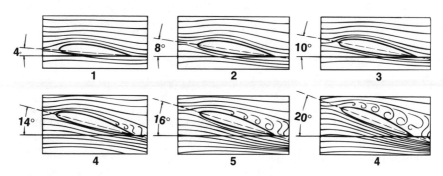

Fig. 3-4 *Increasing angle of attack.*

increasing angles of attack. As you can see, at some point the smooth flow of air above the wing begins to become turbulent. In order for the wing to generate adequate lift, the flow of air must remain relatively smooth as it passes around the wing's upper and lower surfaces. At some point as the angle of attack increases, the airflow becomes so turbulent that the wing is no longer able to produce enough lift to keep the airplane flying. This angle of attack is known as the *critical angle of attack*. For most general-aviation aircraft, the critical angle of attack is normally between 15 and 20 degrees angle of attack.

As you slow an airplane, you must increase the angle of attack to continue to generate the same amount of lift. For this reason we needed to increase the pitch of the aircraft as we entered slow flight in our discussions in that section. While the plane slowed, we were compensating for the loss of lift by increasing the angle of attack. The critical angle of attack is exceeded when we continue to increase the pitch past the point where airflow over the wing can no longer smoothly flow around it. This is when the plane stalls. We are forewarned that a stall is approaching by the rough flow of air over the wings and around the plane. Known as *buffet*, the airplane vibrates and shakes as the turbulence from the rough airflow becomes stronger.

In Chapter 1 we reviewed how pressure differential lift is generated. To summarize, the airflow over the top of the wing travels at a higher speed than the air flowing under the wing. This results in lower pressure above the wing as compared to below the wing. When the wing actually stalls, the airflow has become turbulent to the point that the pressure differential between the upper and lower surfaces of the wing is reduced and insufficient lift is created to keep the plane flying. Figure 3-5 shows the stall patterns on several different wing planforms. You will notice that a wing does not stall all at once. Instead, the stall travels across the surface of the wing. The illustration demonstrates that the propagation of the stall can begin in different spots of the wing and move across in different patterns based on the shape of the wing.

Aircraft designers use this principle to their advantage when designing the characteristics of a wing. For instance, it is preferred that the ailerons remain effective as long as possible to provide roll control. If the ailerons stalled before the rest of the wing, it would be more difficult to maintain a wings-level attitude. Several of the wing planforms demonstrate characteristics that result in the ailerons stalling early in the progression of the stall.

In the remainder of this chapter we are going to discuss a number of different stall scenarios. We will also focus on correct use of controls during stall entry and recovery. We will begin with a comparison of normal versus accelerated stalls.

NORMAL VERSUS ACCELERATED STALLS

So far we have spent a great deal of time covering stalls entered from a slow flight mode. In reality, a plane can stall at any airspeed. A normal stall is usually entered by slowing the airplane to the point that it stalls at a slow airspeed. But if you have flown on a hot summer day, when the turbulence is banging against the plane, and heard the stall warning horn go off as a result of turbulence, you have encountered a situation that could

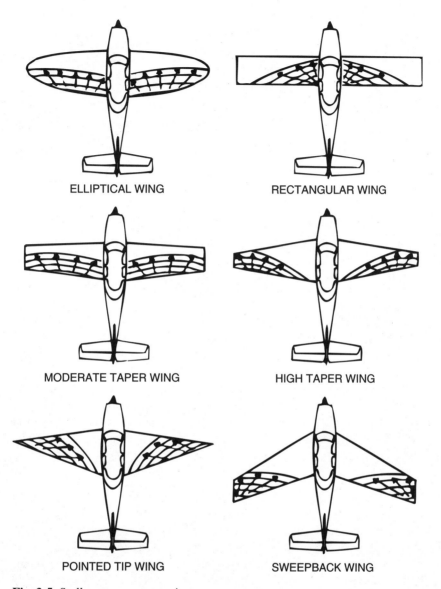

ELLIPTICAL WING   RECTANGULAR WING

MODERATE TAPER WING   HIGH TAPER WING

POINTED TIP WING   SWEEPBACK WING

Fig. 3-5 *Stall patterns across wings.*

result in a stall. If you have put the airplane into a turn and pulled back hard on the yoke and felt the plane shutter, this is also an indicator of a high-speed, or accelerated, stall. Figure 3-6 illustrates a plane flying along at cruise speeds in level flight, the horizontal arrow indicating the direction of flight. The second arrow represents a gust of wind from turbulence. This wind gust can result in an angle of attack that briefly exceeds the critical angle of attack, causing part or all of the wing to stall. Let's take a look at both normal and accelerated stalls in more detail.

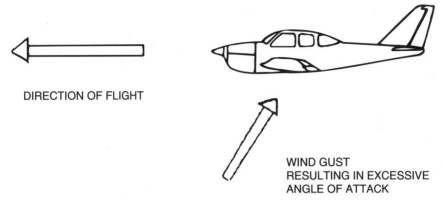

DIRECTION OF FLIGHT

WIND GUST
RESULTING IN EXCESSIVE
ANGLE OF ATTACK

Fig. 3-6 *Gust induced stall.*

Normal Stalls

For the purposes of this chapter we will consider a normal stall one that does not require abrupt control inputs to initiate the stall. For example, a normal stall would be a stall where you have slowed the plane in the manner we discussed for slow flight entry, but continued to ease back on the control yoke to dissipate more airspeed. Eventually the airflow will become so turbulent that the wing will stall and the plane will not be able to maintain altitude.

In contrast, an accelerated stall normally takes place at a higher airspeed, with rapid or excessive control inputs initiating the stall. The airplane is normally under higher g-loads and stalls more abruptly. Accelerated stalls are usually the result of a need to maneuver quickly. As we discuss normal and accelerated stalls, keep this difference in mind.

Normal Stall Entry

We will look at power-on and power-off stall entry as part of our review of normal stalls. You will notice as you read that these two stalls are similar in almost every respect except the amount of power the engine is generating. Entry into both of these stalls is accomplished through smooth, controlled inputs on the controls. How the plane reacts when the stall takes place will vary from plane to plane, but your recovery control inputs should also be controlled and smooth. The airplane should not achieve excessive nose-up attitudes and should remain controllable throughout the maneuver.

Normal Stalls: Power Off

Entry and Recovery. Before you practice any flight maneuver, be sure to execute clearing turns to ensure that the airspace you are flying in is clear. You will not only want to check for traffic at your altitude, but also above and below your altitude. You may have a tendency to gain or lose altitude as you practice, and it is important to verify that no other airplanes are in your vicinity. After reaching a safe altitude and clearing the area, slowly ease the throttle to idle. The reason we do this slowly is to prevent rapid cooling of the engine. After a long climb to altitude, the engine has heated up as a result of generating more power and the reduced airflow around it due to the increased angle of climb. If the engine is cooled too rapidly, this can result in excessive engine wear or damage to engine components.

You will start the maneuver at a specific altitude. Maintain that altitude as the plane slows by easing back on the control yoke, just as you did during slow flight practice. But rather than maintain a given airspeed, you will want to bleed off speed until the airplane stalls. When the stall takes place, you should have the control yoke at or near its aft stop. As the stall approaches you should notice the same sensations we discussed in the slow flight section: soft control feel, reduced engine noise, and the sound of airflow around the plane will become quieter.

Most planes experience a certain amount of buffet as the plane nears stall speed, but the amount will vary for each plane. You may clearly feel the buffet with some planes, while with others it may be difficult to feel. As the plane stalls it will normally experience a nose-down movement known as the "break." The break signals that the stall has occurred and is a pitch-down movement. The amount of break will also be different from plane to plane. In some cases the break can be sharp and very noticeable, while in others the plane may wallow along without a significant pitch down.

While the plane is approaching the stall, or as it stalls, it may have a tendency to drop off on a wing. This can be the result of how the airplane is rigged, or minor differences in the plane's wings' angles of attack. For many pilots the first response is to use the ailerons to maintain wings level during the stall. This should be avoided, instead using the rudder to keep the wings level. Ailerons generate their roll ability by changing the shape of the airfoil. When a plane stalls, this can result in a deeper stall of a wing and a more aggravated stall situation. In some cases this can cause the plane to enter a spin. The use of rudders to hold the wings level as the plane stalls helps prevent this from taking place. If the left wing of the plane drops, you will want to use right rudder to help lift it back up. If the right wing drops, use left rudder to keep it level. Make sure you neutralize the rudder once the wings are level again, and don't overcontrol the amount of rudder you input. Practice with a plane will help you get used to how much rudder should be used as you fly.

At a flight instructor clinic I attended they showed a video clip of a pilot having engine problems. The footage was shot from a boat on a lake surrounded by trees. The pilot was obviously trying to get the plane to the lake for a water landing to avoid impacting trees and was flying just above a stall. The plane would drop off one wing, then the other. Each time the pilot used rudder to keep the wings as level as possible and just cleared the trees, making it to the water. This pilot showed very good use of controls and judgment in avoiding the use of ailerons in that very tense situation. It is quite possible that the plane might have spun in if excessive ailerons had been used in that emergency situation. For me this was a very graphic example of the correct use of rudder in a stall.

As the plane stalls, you will initiate stall recovery procedures. For normal stalls this will include relaxing backpressure on the control yoke to reduce the angle of attack and adding full power. Reducing elevator backpressure will result in a nose-down attitude, which puts the angle of attack less than the critical angle of attack. In most cases you only need to drop the nose of the plane slightly below the horizon. Figure 3-7 illustrates the

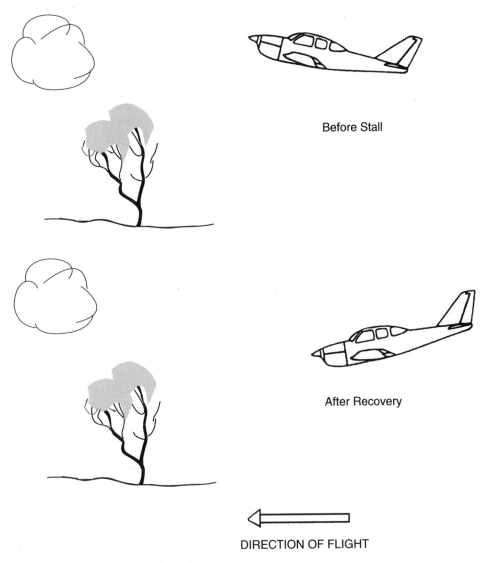

Before Stall

After Recovery

DIRECTION OF FLIGHT

Fig. 3-7 *Power-off stall: pitch angles.*

position of the nose just prior to stall entry and as the plane recovers from the stall. In this case the nose of the plane has been lowered only a small distance below the horizon.

The use of engine power will help the plane recover more quickly from the stall and reduce the amount of altitude the plane loses. Make sure you use full power when you apply the throttle. Some pilots have a habit of easing in only a small amount of power, which results in greater altitude loss. As the plane achieves flying speed, ease the elevator back to produce a positive rate of climb. Avoid pulling the nose up too quickly; this could result in a secondary stall and the need to recover from the stall again with an associated altitude loss. Getting in the habit of correct use of controls and avoiding a secondary stall could mean being able to recover in an accidental stall during an approach to landing.

Common Errors. Students tend to make some common errors during power-off stall practice. One of the most common is not causing the plane to stall completely. This is normally the result of a timid use of the elevator and not getting it back quickly enough, or all the way to the back stop. The results will vary, but the plane may wallow along, not getting a clean break from a good, solid stall. The use of ailerons to hold the wings level during the stall is also common as pilots begin to learn this maneuver. They have been trained to use the ailerons to control the roll of the plane, and now they must learn to use the rudder in place of the ailerons. I have had students ask why this is so important. They say that they have used ailerons and not caused the stall to aggravate further. While in some cases you may be able to use ailerons and get away with it, it is the wrong technique to use for most airplanes. Unless the manufacturer recommends otherwise, use the rudders to control the wings during a stall. An unexpected stall at low altitude is not the place to learn how improper control use can cause additional control problems. Pilots will also forget to maintain right rudder to keep the ball centered as the plane slows, resulting in the plane drifting left. Keeping the ball centered also helps prevent an inadvertent spin.

As the plane stalls, some pilots will shove forward on the control yoke too aggressively, resulting in a significant nose-down attitude. Not only does this pin you against the seat belt and cause all the dirt to float up from the floor, but it also results in excessive altitude loss. In most cases gently releasing backpressure and lowering the nose to just below the horizon is sufficient to break the stall and let the wings start to fly again. Regarding the use of the engine during power-off stall recovery, pilots may forget to add engine power and only use partial power as they recover from the stall. Always add full power in a smooth, constant application. Finally, some pilots will then use too much elevator or apply it too rapidly as they attempt to establish a positive rate of climb. As we discussed, this can cause the airplane to stall again and result in even more altitude loss. Smoothly ease the nose of the plane up as flying speed is increased until the descent is stopped.

At this point you should have an understanding of how to enter a power-off stall, how the controls should be used, and how to recover from the stall. We also covered some of the common errors pilots make and what can happen if you use the controls incorrectly. Now let's move on to power-on stalls.

Normal Stalls: Power On

Entry and Recovery. Power-on stalls are very similar to the power-off stalls we just covered. Flight control use is the same to enter and recover from both stalls, the major difference being that full power is being produced by the engine during entry into a power-on stall. I like teaching power-on stalls to help pilots understand that a plane can stall even when full throttle is applied, and to let them see how high the nose can rise during entry into the stall. Unless they practice them on a regular basis, many pilots are uncomfortable with the pitch angle that is achieved during a power-on stall. But it is important to be aware of what these stalls are like, not only to be able to recover successfully from them, but also to be aware of them and be able to avoid accidentally entering them.

There are a number of different ways to enter a power-on stall, but I like to teach my students to enter from cruise power settings. First ease the nose of the plane up, letting airspeed bleed off to the best angle of climb speed. At that point advance the throttle smoothly to full power. Keep pulling back on the control yoke in a firm, constant motion to keep the nose of the plane coming up and bleeding off airspeed. As with the power-off stall, as the plane slows you will notice that the flight controls become less effective and the airflow around the plane slows. The pitch of the engine will also have a tendency to change as the plane slows and different loads are placed on it as a result of the changing airspeed and pitch angles.

The plane will begin to buffet, as with the power-off stall. My experience has been that planes tend to buffet more in power-on stalls than they do in power-off stall scenarios. The additional airflow generated by the propeller tends to increase the amount of air flowing around the wings and fuselage, and this causes a stronger buffet. The strength of the buffet will be different in each plane and may be more pronounced in some models than others.

As the plane stalls it will once again have a tendency to pitch nose down. I have also found that this nose-down pitch is likely to be stronger than the break we discussed for power-off stalls. As the plane stalls, release backpressure on the elevators and allow the nose to drop just below the horizon to reduce the angle of attack to below the critical angle of attack. The amount of nose-down attitude you will need to achieve to break the stall will differ for each plane, and some planes may require a greater pitch-down attitude than others. The idea is to learn what is correct for each airplane and use the minimum nose-down pitch angle to recover from the stall. After you practice stalls and begin to become proficient in their entry and recovery, you will actually be able to feel when the wings start to fly again. During the stall you can feel a certain mushiness, and as the angle of attack is reduced the wings become more solid in their feel. It takes practice and tuning in to the airplane, but it is possible to feel what the plane is telling you.

Since power should already be set at full power, you will not need to advance the throttle during the stall recovery. It is a good idea to keep your free hand on the throttle to make sure it doesn't creep back during the stall. Once your airspeed has increased sufficiently, ease the nose up and establish a positive rate of climb.

Pilots seem to feel less comfortable with the pitch angles the plane achieves during power-on stalls and how much elevator it takes to get many airplanes to stall when the throttle is at full power. Figure 3-8 shows the pitch angles just before a power-on stall, then after the plane stalls. Compare these to the pitch angles in the power-off stall, and you can see how much steeper the angles are. Because the engine is producing full power, it also takes a considerable amount of right rudder to keep the ball centered as the plane's airspeed drops. Because the nose rises so much, it is often difficult to see the horizon over the nose of the plane, adding to the disorientation that is common for pilots learning the maneuver. All of these factors can be overcome with practice, which helps make you a safer, more aware pilot.

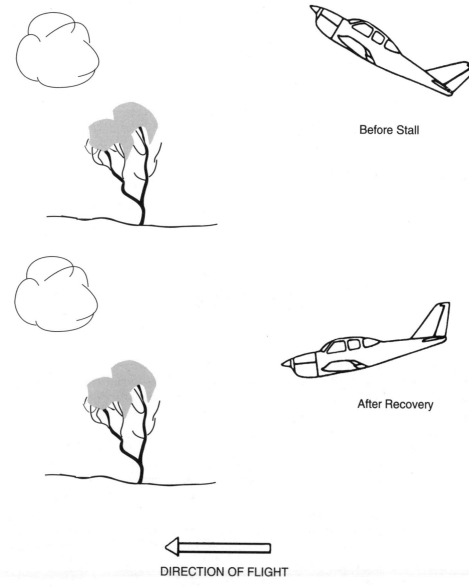

Before Stall

After Recovery

DIRECTION OF FLIGHT

Fig. 3-8 *Power-on stall: pitch angles.*

Common Errors.   Many of the errors we discussed in power-off stalls apply to power-on stalls as well. Remember to use the rudder to hold the wings level during a power-on stall. If a plane has a tendency to drop a wing during power-off stalls, you may find that it is more exaggerated in the power-on stall configuration. Many pilots tend to respond to this sharper reaction from the plane by using ailerons, instead of the rudder, to keep the wings level. Remember to use the rudder to hold the wings level as the plane stalls. Pilots also are not used to the amount of right rudder they need to use to keep the ball centered when the plane is flying slowly with the engine producing full power. It can take quite a bit of rudder to hold the ball in place.

The nose of the plane can drop more rapidly due to the sharper break during the stall and may have a tendency to sink lower below the horizon. Try not to aggravate the situation by adding too much forward elevator input, which can drop the nose even further below the horizon. Every plane will react differently during a stall, but stall recovery is a task that requires that you are light and smooth with flight control inputs. You will also want to make sure you do not set yourself up for a secondary stall by raising the nose too abruptly after your initial stall recovery. The airspeed will have a tendency to build more quickly since the power was on during the entire stall and there can be greater altitude loss as a result of this.

Now that we have covered normal stalls, let's take a look at accelerated stalls. There are many similarities between normal and accelerated stalls, but accelerated stalls can take place at higher airspeeds and be more abrupt in how the plane reacts during them.

Accelerated Stalls

We now know that an airplane can stall at any airspeed, and we understand what takes place during normal stalls. But what is an accelerated stalls and what causes it? In this section we will discuss those areas and also look at ways to avoid accelerated stalls. If you have ever been in a turn, pulled back a little too hard on the elevator, and felt the plane rumble around you, you have experienced the onset of an accelerated stall. In more severe examples the nose of the plane can break very suddenly as the plane stalls. A wing may also have an stronger tendency to drop off during an accelerated stall. An accelerated stall takes place when the plane is placed under g-loads that are in excess of the lift the plane is able to produce at that time. In addition, the critical angle of attack is exceeded, just as with a normal stall.

In addition to the more pronounced stall characteristics, an accelerated stall places higher stresses on the plane than a normal stall as a result of the higher g-loads. Aircraft manufacturers define the maneuvering speed, V_a, as the maximum speed the plane should be flown at when flying in rough air. If you are flying at or below the maneuvering speed, the plane will stall before permanent structural damage takes place due to abrupt control inputs. The example in the previous paragraph demonstrates that these sudden, abrupt control inputs can place significant g-load stresses on the plane's airframe. If you have experienced accelerated stalls you should also have noted the g-forces pushing you down into the seat, a definite indicator you can feel that tells you how much force is being applied to the airframe.

If you encounter strong turbulence, you should always slow the plane to at or below maneuvering speed to avoid damaging the airplane. Flying at airspeeds near but below V_a can provide a margin of airspeed to help avoid an accidental stall but still keep stresses on the airplane to a minimum. Flying at airspeeds above V_a in turbulence or when making abrupt control inputs can result in severe damage to the airplane. And keep in mind that fatigue on an airplane is cumulative, so you may be able to "get away with it" at times, but each time weakens the plane to the point that structures may eventually fail even when the plane is being flown within recommended limits. For that reason, practicing accelerated stalls should NEVER be done above V_a airspeeds for the plane you fly.

Figure 3-9 depicts g-loads the plane is placed under for a given bank angle if you maintain a constant altitude. This graph documents that as the angle of bank increases, the g-loads on the plane begin to increase exponentially. At 80-degrees bank the plane is under approximately 5.76 g's, while at 85 degrees bank the g-loads reach 9 g's. This minor bank increase of 5 degrees causes an extremely sharp rise in the stresses placed on the plane.

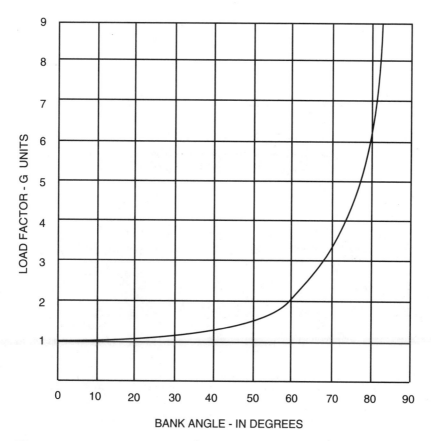

Fig. 3-9 *G-Load versus bank angle.*

Chapter Three

To further understand how accelerated stalls can be a problem if they take place at airspeeds higher than V_a, let's look at Figure 3-10. The airplane you fly is certified under one of three categories. The normal category for certification has a maximum load limit factor of 3.8 g's, which can be exceeded if the plane is placed in a 70-degree bank while maintaining a constant altitude. However, if you are pulling back on the control yoke and banking, the angle of bank needed to exceed 3.8 g's becomes even less. I train pilots in aerobatics, and it is quite possible to pull 4 g's just by pulling back aggressively on the elevator from level flight. In an aerobatic category plane designed for those types of loads, that is not a problem, but in your average general-aviation airplane, this type of abuse can cause severe damage.

You may be wondering why load factors increase in a turn. Figure 3-11 illustrates the forces that act on the plane you are flying while in a turn. In level flight the lift and weight (gravity) are in balance, exactly opposing each other. But as you begin to bank the plane, centrifugal force creates additional loads that are added to those already generated by gravity. Additionally, some of the lift that was used to balance against gravity is now used to turn the plane, so additional lift must be generated to maintain the plane's altitude. This additional lift is created by increasing the angle of attack as you pull back on the control yoke.

Because a plane flying at cruise airspeeds can generate more lift than one flying at slower speeds, the wings are capable of creating enough lift to support weights that

CATEGORY	LIMIT LOAD FACTOR
NORMAL	3.8
UTILITY	4.4
AEROBATIC	6.0

TO LOAD LIMITS GIVEN, A SAFTEY FACTOR OF 50% ADDED

Fig. 3-10 *Aircraft category g-load limits.*

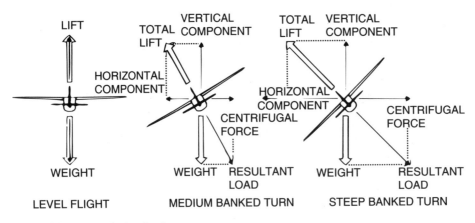

Fig. 3-11 *Vectors during bank.*

72

exceed the structural capabilities of the plane. As you can now see, not only strong turbulence can cause damaging stresses on the plane, but also abrupt use of the flight controls can result in damage to the aircraft. To avoid excessive g-loads, the *Pilot's Handbook of Aeronautical Knowledge* (p. 27) recommends:

1. Operate the airplane in conformance with the *Pilot's Operating Handbook.*
2. Avoid abrupt control usage at high speeds.
3. Reduce speed if turbulence of any great intensity is encountered in flight, or abrupt maneuvers are to be performed.
4. Reduce the weight of the airplane before flight if intensive turbulence or abrupt maneuvering is anticipated.
5. Avoid turns using an angle of bank in excess of 60 degrees.

We have reviewed what an accelerated stall is and what causes it. We have also discussed how to avoid an accelerated stall. But what happens if we encounter one? If the plane does actually enter an accelerated stall, you will want to reduce backpressure on the control yoke, just as in a normal stall. This not only reduces the angle of attack below the critical angle of attack, but also reduces the stresses placed on the airplane. Since accelerated stalls take place in more unusual circumstances as compared to normal stalls, it is difficult to state in advance how much nose-down attitude will be necessary to break the stall but produce the minimum altitude loss. It is also conceivable that the wing will have a tendency to roll off in one direction or another as the plane stalls. Do not overuse the ailerons or rudder to correct for this; it may be best to let the plane remain in a bank as the airspeed builds to safer ranges after you release elevator backpressure. Since you are starting out at higher airspeeds in an accelerated stall, you may have safe flying speed as soon as you release backpressure on the control yoke. If you have sufficient altitude, ease the plane back to straight and level flight after the airspeed is sufficient to allow safe use of flight controls. In some cases you may need to monitor that the airspeed is not building too rapidly. If you find that this is the case, reduce power to help avoid exceeding safe airspeeds.

Common Errors

A number of errors are relatively common among pilots who experience an accelerated stall. The first is not recognizing that they are actually stalling the plane. I have had many pilots ask me what happened after they applied controls too aggressively in an aerobatic maneuver and the plane shook and dropped a wing as it stalled. This seems to be very common when a student is learning to do loops. On the backside of the loop when they are looking straight down at the ground, they have a tendency to start pulling a little too soon and too hard, trying to get the nose back to level flight. All of a sudden the plane starts to buffet noticeably as a result of the load they are placing on the plane, and an accelerated stall develops. Instead of reducing backpressure on the elevators to let the wings start flying again, they keep pulling harder, thinking this will help the situation. To help avoid this situation, recognize that an accelerated stall is taking place and reduce elevator backpressure.

Also, keep the airspeed at or below V_a when you encounter turbulence or perform sudden maneuvers. While this cannot prevent an accelerated stall, it can prevent damage to the plane. Some of the faster single-engine aircraft can cruise at speeds well above the maneuvering speed, and reducing airspeed can be critical for these planes. Knowing the forecast can help you be prepared for these types of conditions.

Avoid steep banked turns, especially in the pattern. While we have discussed accelerated stalls as they relate to higher airspeed, they are just as able to occur at lower airspeeds. If you are in the pattern and attempt to bank too steeply, the plane can stall very quickly. At pattern altitudes you may not have sufficient space to recover before impacting the ground. Be aware of the plane and you can avoid these types of situations.

We have looked at normal stalls and accelerated stalls and discussed each of them in depth. You should now have a good understanding of the difference between these two types of stalls. Now let's take a look at some of the different types of normal stall situations, including approach and departure stalls.

DEPARTURE STALLS

In this section we are going to discuss departure stalls, how they are entered and recovered from, and common errors pilots tend to commit as they practice them. We will also look at how to possibly avoid them in real departure situations, when the plane is at low altitude and you have few options for recovery. In the previous sections we have covered power-on stalls, and the departure stall is merely an extension of that maneuver. We are going to look at how you can add turns into the departure stall to simulate leaving the pattern as you take off and climb out. Keep in mind that you should practice these maneuvers at a safe altitude and that having a flight instructor along can help avoid getting into trouble if you are at all rusty in executing them.

The departure stall is a maneuver that is just what its name implies; the plane stalls during takeoff or departure. The general scenario is that the pilot takes off and does not monitor the airspeed closely enough, letting the plane slow to the point that it enters a stall. Reasons for letting the airspeed get too low vary and can include any number of reasons but might include the pilot becoming distracted by something in the cabin and letting the plane climb at too steep an angle, resulting in a loss of airspeed. The loss of all or partial engine power is also a situation that can cause the pilot to let the plane get too slow. In these situations the pilot may be so focused on the lack of power that he or she does not control airspeed. We will discuss forced landings in Chapter 8, but often the pilot tries to force the plane back to the runway. At low altitudes this is often a poor choice and results in the plane stalling while in a very steep turn low to the ground.

There can be any number of reasons that departure stalls take place, but there is no reason to let the plane stall, even in an emergency situation. Your first duty as a pilot is to fly the airplane. If you control it at all times, you increase your chance of successfully controlling a given situation. That is why practicing departure stalls is so important to recognizing that one is about to take place and being able to correct to avoid it before it happens. Stall recognition and avoidance can reduce the probability that you will need to recover from a stall at low altitude, whether it is a departure stall or an approach stall.

Our discussion of how to enter power-on stalls applies directly to what we are about to discuss. To practice entering a departure stall, you will start at a safe altitude, and be sure to clear the airspace around you before you begin practice. You can pull power back to about 1700 r.p.m. and slow the airplane while you maintain altitude, or you can pull the nose up while power is at cruise settings and slow the plane. Either way, once the plane slows to around V_x, bring in full throttle and begin pulling back further on the control yoke. As we discussed previously in this chapter, the nose will tend to rise quite a bit above the horizon when you are doing stalls with full engine power. Figure 3-12 shows the view looking to the left during a departure stall flying straight ahead. Remember to keep the ball centered as the plane slows.

As the plane approaches stall speed, you may feel a certain amount of buffet. Different planes have a different amount of buffet, from very pronounced to barely noticeable. You will need to become used to the signs the plane you are practicing in gives you. As the plane stalls, keep the wings as level as possible, using rudder to pick up a wing that might drop. Let the nose fall just below the horizon by releasing backpressure on the elevators and as soon as the airspeed is at an acceptable value, ease the nose of the plane back to level flight. You should have already had full power, but make sure the throttle is all the way forward. Figure 3-13 shows a forward and side view from the plane just after the stall takes place and the nose has been lowered. The nose of the plane has not been lowered excessively, which helps reduce the amount of altitude lost during stall recovery.

Once you become comfortable executing departure stalls straight ahead, begin making turns to the left or right as you pull the nose of the plane up in preparation for entering

Fig. 3-12 *Departure stall: left view.*

Fig. 3-13 *Departure stall: recovery forward and side views.*

Fig. 3-14 *Departure stall: forward view.*

the stall. Do not exceed banks in excess of 30 degrees since you should not use a bank steeper than that in the pattern. As with straight-ahead departure stalls, slow the airplane to V_x, then add full power and ease back on the control yoke to begin bleeding off airspeed. Maintain a constant bank as the plane slows; do not let it steepen or shallow. As the plane stalls, release backpressure on the control yoke to let the nose drop to just below the horizon. Figure 3-14 shows the nose of the plane just as the plane stalls. As you can see, in this example it is in approximately a 30-degree bank. Since the plane is now banked, you may find that it stalls at a higher airspeed due to reasons we have previously discussed in this chapter. It is important to keep the ball centered through the course of the departure stall to help prevent an accidental spin.

Use the rudder to keep the plane from dropping a wing further, which can happen as the nose of the plane breaks when it stalls. I like to teach pilots not to worry about getting the wings level immediately, but instead to make sure they get the nose down and regain flying speed. Keeping the bank steady during the stall avoids the tendency to use ailerons to roll the plane level during the stall, which, as we have learned, can aggravate the stall. The highest concern is to get the nose down the right amount, and get the plane flying again. Once you have sufficient airspeed, roll the wings level using ailerons and ease the elevators back to initiate a climb.

Start out with shallow banks and increase their steepness as you become more proficient. You will find that many planes break more sharply during the stall as the angle of bank is increased and may have a tendency to roll off on a wing during the stall. In some cases I have found that the rolling tendency may be in the opposite direction as the initial direction of the bank. For instance, if you are in a turn to the right as the plane stalls, it may roll back to the left. For many pilots there is a very strong tendency to use the ailerons to control the roll when this happens, but you should avoid this first reaction. Let the plane break, but use the rudders to keep the roll from becoming too steep. Then as the airspeed builds you can roll back to wings-level flight and raise the nose to a climb attitude.

Common Errors

Like standard power-on stalls, departure stalls can be intimidating due to the relatively high pitch angle the plane achieves prior to the stall. Pilots are sometimes hesitant to pull back on the control yoke enough to slow the airplane down to stall speeds. In some cases the plane just mushes along, never getting a good break as it stalls. Like all maneuvers, it takes practice to discover how to use the controls to get the airplane to stall properly. Pilots also have a very strong desire to use the ailerons to roll the plane to wings level just as the plane stalls. Make sure you concentrate on using the rudders to control the angle of bank through the stall, and use the ailerons to roll the wings level once you have achieved safe flying speed.

It is not uncommon for pilots to drop the nose too far below the horizon during departure stalls. The higher pitch angles achieved during the stall seem to incline pilots to let the nose drop to higher pitch angles after the stall. While every plane is different, you want to lower the nose just enough to decrease the angle of attack and let the wings begin to fly again. Though it may be tempting to let the nose fall through in an effort to get airspeed back, you are actually losing more altitude than you need to as a result. Once flying airspeed has been established, do not try to pull the nose up with excessive elevator backpressure; as previously stated, this can lead to a secondary stall and additional altitude loss during recovery.

You should not be intimidated by departure stalls. While they should be given the respect any stall is due, they are not beyond the ability of the average pilot to become proficient at. I like unusual attitudes and find that helping other pilots get used to the pitch angles as they practice them can help make them more at ease as they practice. Once you become competent at departure stall entry and recovery, you will find that it will help you avoid entering one accidentally during actual departures. Let's move on to approach-to-landing stalls.

APPROACH STALLS

Approach stalls are very similar to power-off stalls. While in some planes you may carry a certain amount of power during your approach to landing, it is usually a minimal amount compared to departure stalls. In this section we will review how to enter and recover from approach stalls and common errors pilots can make as they practice them. As you practice approach stalls, note the difference in feel and the airplane's reaction during the stall as compared to departure stalls. For most single-engine general-aviation aircraft, the approach stall is more docile, the pitch angles are generally lower, and the break is often less severe. But don't let approach stalls lull you into a sense of complacency; it is just as easy to end up in an uncoordinated control use situation, which can aggravate the stall, possibly resulting in the development of a spin.

Like the power-off stall, pull the power back to idle and set up for approach airspeeds. Remember, use the elevator to control your airspeeds as you practice this maneuver. Since you are starting out gliding, you will want to make sure you have sufficient altitude to allow you to enter the glide, stall the airplane, and recover from the stall. You should practice approach stalls with varying degrees of flaps, from no flaps to full flaps, noting the difference in how the plane reacts with each change in flap settings.

Slow the airplane by easing the control yoke aft. As the plane slows you will begin to notice the same imminent stall warning signs we discussed previously. The controls will become less authoritative, the sounds of airflow will diminish, and the airflow around the airfoil will begin to buffet as the angle of attack increases. As the plane stalls, the nose will have a tendency to break, dropping as the wings finally reach the critical angle of attack. As the nose drops, release backpressure on the elevator control and allow the nose to drop slightly below the horizon. Depending on the plane you fly and the angle of the nose during the descent, you may need to let the nose drop somewhat lower as compared to the power-off stalls we reviewed previously in this chapter. The goal is to reduce the angle of attack just enough to break the stall, but not so much that you lose excessive altitude as you recover. While the nose may be lower in relation to the horizon as compared to a power-off stall done from level flight, it should not be allowed to drop too far.

As you release backpressure you will also want to bring in full power. We know from earlier in this chapter that this helps reduce altitude loss and allows you to establish a positive rate of climb once the airspeed increases sufficiently. If you have flaps extended you will want to get in the habit of retracting them in increments once you establish a positive rate of climb. In a true approach stall scenario, you may be low to the ground and need to climb to avoid obstacles. For many aircraft, having full flaps can reduce its ability to climb, and this may hamper your efforts to climb above obstacles. Ideally you will configure the plane according to the aircraft manufacturer's recommendation for best angle of climb, or for a go-around situation. The aircraft's operations manual should be consulted for the proper recovery procedures. Flaps should be retracted in increments, allowing the airspeed to build sufficiently to prevent the airplane from sinking as you retract the next notch of flaps.

Like departure stalls, you should practice approach stalls straight ahead until you become comfortable with them, then move on to approach stalls in a turn. You may find it is difficult to get a good, clean break during departure stalls. I have had cases during training where the plane just mushes along, even when the elevator control is all the way back. This mushing tendency has sometimes been overcome by attempting the departure stall again, with more aggressive aft elevator input. Every plane will react differently, so it is difficult to predict exactly how the plane you are flying will react. Through practice you will soon learn how it flies in these stall situations, and the best methods for recovery. It might also be a good idea to use approach power settings during practice, becoming familiar with how the plane stalls when partial power is carried during the approach. The nose will be higher during these types of stalls, and the break may be more pronounced, but stalls with partial power should still be less abrupt than departure stalls.

Common Errors

Like each of the stalls we have reviewed so far, there are a number of errors that pilots seem to make as they learn or reacquaint themselves with power-off stalls. Correct use of rudder is very important, but you may find that less right rudder is needed to keep the ball centered than you use during a departure stall. Remember to apply full power as the plane stalls and transition to a climb after you have attained safe flying speeds. Pilots will sometimes force the nose of the plane up too quickly. In a true low-level stall, you do not want to risk a secondary stall close to the ground. Learning the minimum that you need to lower the nose to break the stall, allowing airspeed to build, and then beginning the climb will help you avoid secondary stalls. Some pilots retract the flaps all at once during stall recovery. For many airplanes this will cause the plane to sink due to the loss of lift until the airspeed increases, resulting in greater lift and the ability to arrest the rate of descent. In a real emergency, you don't want to have recovered from the stall and avoided the ground, only to sink into it by retracting the flaps too quickly. In addition, appropriate use of ailerons and rudder are important, as we have already covered.

STALL AVOIDANCE

So far this chapter has covered how to practice a number of different stalls. Practicing stalls at safe altitudes is a great way to learn about stalls and how the plane reacts, but in this section we are going to discuss how to avoid accidental stalls. We will review common errors that can lead to accidental stalls, imminent stall warning signs, and your responsibilities as a pilot in accidental stall avoidance. Safety should always be the uppermost concern when you fly, and learning how to recognize an approaching stall and how to prevent one from taking place can be the best action you might ever take as a pilot.

To learn to avoid stalls, to get to the point that you can recognize them instinctively, you must first be very familiar with stalls. That is precisely the reason we have spent so much time discussing the various types of stalls you might encounter: normal, accelerated, approach, and departure are all stalls that you should be very comfortable with. Once you can enter and recover from these stalls, and as you practice each of these maneuvers, you gain a feel for them. This feel, the ability to know what the stalls make the plane's control surfaces feel like, the attitudes and sounds of the engine, and slipstream are the core of being able to recognize the onset of a potential stall. Through practice you can not only get better at entry and recovery from stalls, but you also gain the insight to know when the airplane is about to stall, without looking at the instruments. This knowledge is crucial to avoiding stalls when situations become critical. So go out and practice stalls as often as you can; it will help you become more proficient at stall avoidance as well.

As you are practicing, what should you be paying attention to that can help in learning stall avoidance? There are a number of warning signs, many of which we have already covered. The flight instruments can provide a great deal of information via the airspeed indicator and the turn and bank indicator. If you notice the airspeed indicator dropping toward the bottom of the green arc, you know the airplane is becoming a little too slow. If the artificial horizon shows an abnormally large pitch, it might be a good idea to check the airspeed and verify that you are not bleeding off too much airspeed.

Most stalls are not caused when the pilot is closely monitoring the instruments, though. It is not possible to document each potential accidental stall scenario, but some common factors should be considered important. Often the pilot is distracted in one way or another and does not pay attention to the plane. This distraction can come in the form of trying to force the plane into a steep bank as the pilot attempts to avoid overshooting the turn from base to final. The pilot can become so focused on the turn that he or she loses focus on how steep the bank is in relation to the speed of the plane. An accelerated stall can have very rapid onset, with little time between prestall buffet and the actual stall, making it more difficult to recover once the stalls become imminent. For these reasons it is very important to not get yourself into accelerated stall situations, especially when flying in the pattern. Do not focus so heavily on one aspect of the flight that you ignore other pieces. Keep a broad view of what is going on; get into the habit of being "situationally aware" as you fly.

Yet another common distraction is that something happens in the cockpit to take you away from paying attention to flying the plane. This type of distraction can include radio chatter or the tower asking you to do something, looking for an approach chart or sectional, a passenger asking you questions as you get ready to land, or any number of other things that can pop up during a flight. Emergencies are a great way to become distracted. An engine that suddenly begins to run rough and loses power can cause you to focus almost exclusively on the oil and fuel pressure gauges if you don't make yourself stay aware of the entire situation. Gear lights not showing green as you get ready to land can also be a situation that might have you looking at the lights, head buried in the cockpit. In a case like this you are already slow, head looking down at the gear indicator lights, and if you lean forward to tap on them, you may unknowingly pull back on the control yoke as a brace. You probably have heard lots of hanger flying stories in your aviation career from pilots who are not included as examples here. Just know that you will someday encounter a situation that is going to tempt you to focus on a very narrow aspect of your flight, but make sure you keep your focus on the entire flight, including flying the airplane. If you don't fly it, the plane will not fly itself.

Other factors you should recognize are also very important in accidental stall avoidance, many of which we have already covered but are worth reviewing again. Feel the controls as you fly. If you have your head in the cockpit, you should still be able to notice when the controls become softer as the slipstream over them weakens due to dropping airspeeds. This is a definite indicator that you are getting slower than you want to, especially if you are in the pattern or at low altitudes.

Sounds are the other major sensory input that can help you maintain situational awareness. As the plane slows, the sounds generated by the slipstream will lessen and become quieter. Like the feel of the controls, you do not need to be looking outside the plane to notice this change in sound levels. Conversely, if the plane is picking up speed, the noise of the slipstream will increase. This situation will not normally result in a stall, but if you are already at pattern altitudes and you accidentally drop the nose and the plane is losing altitude and picking up airspeed, this can give you a sensory queue as input. Engine noises will also change with changes in airspeed, and any difference in sounds from the engine can also be an indicator that your airspeed is changing either up or down. Listen as you fly and learn to judge the plane's airspeed by the sounds you hear, then compare that to what the airspeed indicator is telling you. You'll find that you can become fairly accurate at "guesstimating" the airspeed based on the sounds you get from the plane.

Pitch angle can also be another indicator that you can use to potentially judge your airspeed, but if you are distracted by something in the cockpit, this may be something you would notice after the feel of the control and sounds lead you to guess you are slowing down. I find that pitch angle is useful during the takeoff phase, though. During takeoff you should be familiar with the normal pitch angles the plane you fly uses to achieve airspeeds such as V_x and V_y. If you begin to notice that the nose is higher than normal during takeoff, you should verify your airspeed. In some cases this pitch angle/airspeed difference can be caused by improper use of flaps, the gear being down/up when it should be in the opposite position, or improper propeller or engine power settings. Obviously you will need to be looking outside the plane to notice excessive or shallow pitch angles, but this can be a clue that other factors are affecting the performance of the plane. High angles of attack can lead to low-altitude stalls, so be very careful to avoid getting yourself into that particular corner. If you notice you are excessively nose high but are low to the ground, avoid dropping the nose too rapidly, which could cause the plane to lose altitude, possibly striking the ground. In that scenario you will need to use judgment and finesse to avoid an accident.

Common Errors

Since every accidental stall situation can be unique, it is difficult to pinpoint common errors that pilots may have committed in not noticing that they were approaching a stall. The most likely error pilots that have accidentally stalled can commit is to not pay attention to flying the airplane. We have already covered a number of potential scenarios that might distract a pilot, but discipline and focusing on flying the airplane are the best methods to avoid letting the airplane get away from you. As the pilot, safety is your highest priority, and you should fly the airplane at all times; do not let the airplane fly you.

Another common error in accidental stall situations is to take too long to recognize the stall and recover once the stall has taken place. With practice you can learn to lose very little altitude when you execute planned stalls, and with enough practice recovery from them becomes second nature. This is the type of automatic reaction you want to build so that if an accidental stall occurs, training takes over and you recognize and recover from the stall as quickly as possible. At low altitudes you have very little time to make the recovery, and being able to react can make the difference in how successful you are.

The last common error we will cover is overreacting during the recovery and causing a secondary stall. As I fly airshows, it is not uncommon to be looking straight down at the runway from several hundred feet altitude, and maintaining focus is very important to knowing what attitude the plane is in and how to successfully complete the maneuver. But I am expecting to see the view from the cockpit and anticipate what I need to do. If you are caught in a nose-down attitude unexpectedly as the result of a stall, looking at the ground can cause you to pull back excessively on the elevator control, preventing recovery from the original stall or causing you to enter a secondary stall. In either case this can prevent a successful recovery from the stall. While it may be tempting to pull hard and fast on the elevators, use finesse on the controls as you fly. If you practice flying stalls—making a conscious effort to use light, coordinated inputs on the controls—you will be less likely to overcontrol the plane in an accidental stall.

CONCLUSION

We have covered a great deal of information related to slow flight and stalls in this chapter, from what causes left-turning tendencies during slow flight to normal and accelerated stalls and then on to departure and approach stalls. Practice of each of these maneuvers has been stressed throughout the chapter, with safety as your number one priority as you practice. Many pilots I have flown with rarely, if ever, practice slow flight or stalls outside of the biennial flight review, and some have not even been taught how to execute full stalls during their flight training for private pilot ratings. This lack of familiarity with slow flight and stalls makes them wary of how the plane will react during a full stall. As a result they are unprepared to be able to recognize the onset of a stall, or to be able to automatically recover if an accidental stall takes place. The NTSB has documented far too many low-altitude stall/spin accidents that may have been preventable if the pilot had been aware of the potential for a stall and avoided entering one.

You should not be afraid of stalls; they are a normal part of flight and can help you learn how to fly the plane in a controlled and safe manner as you recover from them. The biggest problems that pilots have expressed concern about regarding stalls is that they were never properly trained, or it has been so long since they have made an effort to practice stalls that they are rusty in those skills. If you fit into this category, don't feel alone. The best way to remedy the situation is to find a qualified flight instructor and go out and practice slow flight and stalls until you feel comfortable executing them. While not everyone will learn to like stalls, you can become proficient and comfortable with the entry into and exit from stalls. Practicing can also help you learn to recognize the warning signs that a stall is imminent and help you avoid accidental stalls, which is the primary purpose of this entire chapter. The stalls we execute on purpose are not the ones that normally cause problems for us; it's the ones that we don't expect as we turn from base to final that can bite hard. Knowing what causes a stall, how factors such as steep turns can increase the stall speed of the plane, and the indications the plane can give you to warn of a stall can mean the difference between avoiding or not avoiding an accidental stall. You are the pilot and your passengers depend on you to know how to keep them out of trouble as you fly. That alone is reason enough to take the time to learn the proper way to execute and recover from a stall.

Chapter Three

In the next chapter we will look at spins, which are closely related to stalls. As you will see, you must stall a plane before you can spin it. Like this chapter, the main purpose of Chapter 4 is to help you learn about spins so that you can recognize what causes them and avoid accidental spins. You will see what quickly becomes clear—that if you don't stall, you can't accidentally spin in, and the stall/spin accident is another common accident situation that needs to be avoided. As you read Chapter 4, many of the concepts that we discussed in this chapter will continue to be developed further.

4
Spins

Spins are not required for private or commercial flight training, but they are such an important skill that I felt it was mandatory to include a chapter covering their entry and recovery. Like stalls, knowing what causes a spin, how to enter it, and how to recover from it can help you learn how to avoid accidental entry into one. Should you happen to accidentally enter a spin, knowing how to recover from it quickly and with a minimum of altitude loss can be extremely valuable. I actually like doing spins a great deal, practicing them on almost every flight as I give aerobatic training. Prior to getting into teaching aerobatics, I taught private-pilot students spin entry and recovery as a matter of course.

You may be asking yourself why spin training is important if the FAA does not consider it a skill you need for private or commercial flight training, and why it deserves its own chapter in this book. Like accidents involving stalls, spins appear all too frequently in NTSB accident reports. Because pilots may not have received proper training regarding spins, they are not really sure what causes them, what a spin feels like, or how to quickly recover from one. If this chapter convinces just one reader to go out and get spin training, then it is worth including. I don't even expect everyone to like spins as much as I do, but knowing how they feel, how to use the controls, and what you need to do to recover is enough. Learning to like doing spins is just the frosting on the cake.

I'll never forget the first spin I did; it was in a J-3 Cub. I remember watching the nose rise as the plane stalled, then felt the rudder push to one side and the plane's nose dropped what seemed to be almost straight down. With a huge smile on my face, I looked ahead as I watched the ground spin around in front of me. I was hooked; I had to learn how to do a spin on my own. I didn't get to learn then, but later during my flight-instructor training I learned to execute spins in a Citabria and thought it was the greatest thing in the world. I'd like everyone to feel that way about spins, but since that is not likely I just want everyone to understand spins and be able to recover from them safely. Armed with this knowledge, you are better equipped to avoid accidental spins and become a safer pilot.

Safety should always be the primary concern when you fly, and as you begin learning spins from a qualified flight instructor in a plane certified for spins, keep this in mind. Find a flight instructor who is competent in spin training and who has a plane certified for spins. Heading out to the airport and jumping into a plane that is not certified for spins can make for the type of excitement you want to avoid. Before you begin spin training, talk with the flight instructor and get a feel for his or her spin training experience. Then check out the airplane the instructor wants you to use. Many aircraft are not certified for spins; the certification is based on passing specific tests laid down by the FAA. If you don't feel comfortable with what you see and hear, find another flight school that will meet your needs. Let's begin the chapter with a look at what a spin is.

SPIN DEFINITION

I have heard spins described in various ways, often with expletives scattered amongst the description. In this section we are going to define what a spin is by describing what takes place in a spin. As you will see, there is no mystery. Planes enter spins for very specific reasons. For a plane to spin, first one or both wings must have stalled. Second, you must have a situation where one wing has a greater angle of attack than the other. The wing with the lower angle of attack is able to produce more lift than the wing at a higher angle of attack, resulting in a difference in the amount of lift produced by the wings. (Remember that the wing that has the higher angle of attack is past the critical angle of attack and is producing less lift as a result.) The different angles of attack for the wings can be the result of a number of factors that include how the plane is rigged, whether the flight controls are being used in a coordinated manner, or other factors.

Whatever the reason, the stalled plane then begins to yaw in the direction of the wing that is more stalled and the plane begins to slip in the direction it is beginning to yaw. As the plane yaws, the situation becomes worse as the side of the fuselage, the vertical stabilizer, and other vertical surfaces weathervane into the wind (*Flight Training Handbook*, p. 156). Figure 4-1 depicts the differences in the angles of attack between the wings as the plane enters a spin. The asymmetrical lift also results in a rolling moment, causing the plane to roll in the direction of the stalled wing. Due to the yawing, rolling, drag, and centrifugal forces at work, the plane continues to pitch nose down, rolling and yawing, also known as a spin. Each plane reacts differently during the course of a spin; the same

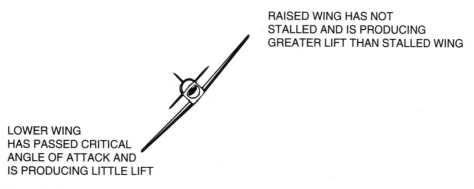

RAISED WING HAS NOT
STALLED AND IS PRODUCING
GREATER LIFT THAN STALLED WING

LOWER WING
HAS PASSED CRITICAL
ANGLE OF ATTACK AND
IS PRODUCING LITTLE LIFT

Fig. 4-1 *Spin angle of attack.*

plane can even exhibit different characteristics, depending on how it is loaded and other factors. As you will discover later in this chapter, the pitch angles and rate of rotation during a spin will vary as the spin progresses.

Once a plane enters a spin, it will remain there until recovery is initiated. If you are flying a plane certified for spins, multiple-turn spins are safe as long as you remain within the limits specified by the manufacturer. A note of warning, though—never purposely spin a plane that is not certified by the manufacturer for spins. For those aircraft certified in the normal category, the manufacturer is only required to demonstrate that the plane is able to recover from a one-turn spin, or three second spin, whichever takes longer. If you spin a plane in this category, once the plane passes one turn, you have become a test pilot, because the flight characteristics are unknown. In some cases the plane may enter a spin mode that you may not be able to recover from. This is the reason I stress practicing spins only in spin-certified aircraft.

Stall/Spin Relationship

We learned what a spin is in the previous few paragraphs—a situation where one wing has stalled and the other is producing lift, which results in a yawing, rolling series of motions. But what starts the entry to the spin? Can you just be flying along when the plane suddenly enters a spin, or be in a steep turn? Basically, to spin the airplane you must first stall it. In Chapter 3 we learned that stalls can take place at any airspeed, so it is conceivably possible to spin an airplane at any airspeed. In fact, an aerobatic maneuver known as the snap roll is a high-speed stall/spin, resulting in the plane spinning in the direction the plane is moving. This can be going horizontally, vertically, or straight down and is a very rapid, quick maneuver. You definitely do not want to attempt this type of spin in anything but an aerobatic plane capable of doing snap rolls, as any other plane may have the wings depart the fuselage during the snap.

You will also want to consider that if a plane is difficult to get into a spin, it may also be difficult to get out of the spin. While this is not always the case, be certain you know the airplane you are practicing spins in and the correct spin entry and recovery procedures. We will discuss general spin entry and recovery procedures in this chapter, but the plane you are flying may have a unique set of techniques that are documented in the aircraft operations manual.

Spin Versus Spiral

I have flown with pilots who are confused regarding the difference between spins and spirals. From the cockpit these two maneuvers can appear very similar; the plane is in a nose-down attitude rotating to the left or right. There is, however, a very large difference between spins and spirals. During a spin the plane stalls and as a result of asymmetrical lift enters the rolling, yawing actions we have discussed. At least one wing is stalled and the other is producing more lift. The airspeed indicator will not be increasing and stays near the speed of the stall.

A spiral can be entered from a botched attempt to spin the plane, but a spiral is significantly different in what is aerodynamically taking place. Unlike the stall, the wings are not stalled in a spiral; they are both producing lift. The rolling moment found in a spiral is normally due to aileron and/or rudder control inputs. Another major difference that you can actually verify is that the airspeed will be increasing in a spiral. If you are in a spiral, the plane is actually in a dive and will pick up speed in the dive. This can become a dangerous situation if you do not recognize the difference and let the airspeed continue to increase. If the plane accelerates to speeds beyond V_{ne}, you could find that structural damage can cause significant structural failure.

Recovery from spins and spirals also differ. We will discuss spin recovery in more detail in this chapter, but essentially you want to pull back the power on the engine to idle, input full rudder in the direction opposite to the direction of spin, and ease forward on the elevator control to reduce the angle of attack and break the stall. To recover from a spiral, you also want to reduce engine power to prevent excessive speed buildup, but you will not need full rudder opposite the direction of the spiral's rotation and you will not want to push the nose of the plane down further through forward elevator control inputs. This would actually result in the plane entering a greater nose-down attitude and gaining even more airspeed. The use of full rudder opposite the direction of the spiral could actually cause excessive yawing and additional controllability problems.

If you do enter a spiral, you will want to first reduce engine power to keep the plane from gaining speed too quickly. You will then want to ease the nose of the plane back to a level flight attitude. Avoid pulling back too rapidly on the elevator control; excessive use could cause a high-speed stall situation, further compounding problems. Too much elevator too quickly could also place excessive g-forces on the structure of the plane and result in damage to the aircraft. Finally, use ailerons to roll the wings to a level flight attitude. Like the elevators, do not use the ailerons to excess as you stop the rolling tendency of the plane.

Spiral entry will not happen the same way for every airplane. You may accidentally roll the plane into a spiral or find some other unique manner for putting the plane into a spiral. On the occasions I have entered a spiral, it has been the result of attempting to spin a plane that is reluctant to do so, and the aircraft rolled off into a spiral. I had full rudder in the direction of the spin I wanted to fly, had the elevator fully aft and power off, and the plane just did not want to spin but would instead drop the nose down and start spiraling downward, picking up speed. Some planes are like this, while others would rather spin than spiral. Experience is the best teacher for getting a feel for what a spin feels like, as compared to a spiral.

SPIN PHASES

A spin is not a static maneuver; instead the plane moves through several distinct phases with the characteristics of the spin changing in each phase. In this section we are going to discuss each of those four phases: spin entry, incipient spin, developed spin, and spin recovery. Other books may use other names, but the basic concepts documented during each phase of the spin are still the same.

Spin Entry Phase

We have already touched on the spin entry phase. As discussed previously in this chapter, the spin begins as the plane stalls and the rolling/yawing motions begin. The nose of the plane normally drops well below the horizon as the spin begins to take effect. To enter a spin you will slow the airplane, just as you would with a stall. Be certain to use carb heat if the airplane you are flying needs it before you pull the power back. Ease back on the control yoke to slow the airplane and maintain altitude. As the plane stalls make sure the elevator control is fully aft, assuring a complete stall, then push full rudder in the direction you want to spin the plane. For example, if you want to spin to the right, push full right rudder and hold it there. This will cause the plane to enter the second phase of the spin, the incipient spin phase.

Incipient Spin Phase

The incipient phase of the spin takes place as the plane transitions from forward flight to the nose-down, rolling, yawing descent present in a spin. Inertia, lift, drag, and yaw all affect the plane during the incipient phase. In fact, NASA has expended a great deal of time and resources studying spins and what factors affect them. I have already written a book published by McGraw-Hill, *Spin Management and Recovery*, that goes into great detail regarding spins. We will only enter a high-level discussion of some of the factors affecting spins.

As the plane enters a spin, inertia attempts to keep the plane moving in its original direction. The inertia must be overcome during the incipient phase of the spin to transition the plane from horizontal flight to the nose-down descent found during the spin. Lift, yaw, and drag must also transition to the new direction of the plane, and it takes time for each of these forces to change.

It generally takes approximately two turns for the plane to make the transition and overcome the forces attempting to keep it moving in its original direction of motion. Due to the tendency of a plane to want to continue in its original motion, in many cases it is easier to recover from a spin while the plane is in the incipient phase. In fact, while not an approved method of recovery for most aircraft, letting go of the controls while the plane is in the incipient phase can often result in the plane recovering from the spin on its own. DO NOT assume this is the case for the plane you are flying and go off and practice spins on your own. There are many reasons why this technique may not work and could make for a very exciting day for you in this case. Self-taught spins are not recommended for anyone. At the end of the incipient phase, the plane's momentum has transitioned into a fully developed spin, and it is much less influenced by the forces originally acting on it during spin entry.

Developed Spin Phase

As the spin progresses, the influence of the inertia and other forces is reduced and the plane settles into a developed spin. Spins are generally misunderstood, and this phase presents some interesting aspects when it is studied. Contrary to the belief of many pilots, the developed phase of a spin is not an unchanging situation. The plane actually goes through cyclical periods as part of the developed phase. The changes here include differing pitch attitudes and rates of spin rotation. For example, the Pitts Special S-2B I fly has a tendency to be nose low at the 180-degree point of the spin, and the nose is higher at the 360-degree point. Although this change in pitch angles can be difficult to notice when pilots are first exposed to spins, after they gain experience and comfort it becomes apparent that spins do not remain constant through their course.

Every plane is different in the way it reacts during the developed spin phase. Some airplanes will have a more nose-down tendency during the spin, while others may have faster rotation rates. Depending on how you enter the spin and apply controls, the same plane may behave differently from spin to spin. The center of gravity can also affect the spin characteristics of a plane. Generally, the more aft the center of gravity is, the flatter the spin will be. In our discussions of flat spins later in this chapter, we will take a more in-depth look at this effect.

During spin training I have experienced students abruptly pulling up to enter the stall/spin, mushing the plane into the spin through timid application of controls, and many other interesting applications of the flight controls during spin practice. Even the same aircraft can surprise you in how it reacts to spin entry, and the entry technique can affect how the spin will develop to some degree. Make sure you understand the plane you are flying as you observe what takes place during the developed phase of the spin.

Spin Recovery Phase

We will cover spin recovery techniques throughout this chapter, but at this point we will take our first look at how the FAA recommends that you recover from spins. The recovery phase of the spin takes place as you apply correct flight control inputs that stop the rotation caused by the yawing moment and then break the stall the wings are in. Every plane can react differently than described in this section, the recovery techniques we will discuss are in general what work best for recovery. Before you ever execute spins in an airplane you should review the operations manual for that plane to verify the techniques the manufacturer has determined will work best. You should also be aware that it may take several turns from the point the spin recovery controls are applied before the plane will recover from the spin. It is very important that you know the characteristics of the plane you are flying to know whether it is reacting correctly.

According to the FAA (*Flight Training Handbook*, p. 157), the proper spin recovery technique is as follows:

1. Retard power.
2. Apply opposite rudder to slow rotation.
3. Apply positive forward elevator movement to break stall.
4. Neutralize rudder as spinning stops.
5. Return to level flight.

If the operations manual for the plane you are practicing spins in lists a different set of procedures to recover from the spin, use those rather than those listed here. I have used these techniques in a number of different airplanes, including the Pitts Special, Cessna 152, Citabria, Decathlon, and Piper Cub, and it works well with each of them. Generally, the plane reacts immediately, recovering from the spin in less than one turn if the control inputs are correctly applied.

SPIN ENTRY

The previous section touched briefly on spin entry techniques as part of the discussion of the four phases of a spin. In this section we will look at how to enter a spin in more detail, review what can cause an accidental spin, and discuss common errors pilots make during spin entry. The techniques we will cover here are for entry into normal spins. Later in the chapter we will get into more exotic spins, and entry techniques may differ for those maneuvers.

Normal spin entry is made from a power-off, flaps-retracted stall. Before beginning the maneuver, you should execute clearing turns to the left and right, paying particular attention to the airspace beneath you. Depending on the plane and the number of turns you will make during the spin, you could lose several thousand feet of altitude, so it is important to make sure you do not have any traffic under you that could pose a safety concern. Use carburetor heat as recommended by the plane's manufacturer.

Figure 4-2a shows the attitude of the plane looking over the nose just prior to stall. Figure 4-2b shows the view looking out the left window from the pilot's seat. The angle of attack is very apparent in this view, the angle between the lower surface of the wing and horizon clearly visible. As the plane stalls, the ailerons should be in the neutral position and maintained in that position throughout the course of the spin. We will discuss ailerons and their effect on the spin during the section on flat spins, but you should be aware that if the ailerons are turned into the spin, or away from the spin, they could cause a significant change in the characteristics of the spin. The rudder should be briskly applied in the direction you want the plane to spin, and the elevator should be moved to the full aft position if it is not already there.

Fig. 4-2A *Forward view just before stall.*

Fig. 4-2B *Left view just before stall.*

Figures 4-3A and B through Figure 4-4A and B show the forward and left views of a spin as it progresses. As you can see, the change in pitch as the plane enters the spin is dramatic. The plane used in this photo sequence has a tendency to roll slightly inverted as it first enters the spin, then it settles down into a steep nose-down attitude as the spin progresses. The steep nose-down angles and rapid rate of rotation can be disorienting to many pilots new to spins. I like to count off each rotation to help the student stay oriented. It has been my experience that without this initial help in staying oriented, pilots who have completed only a one-turn spin think they have actually made three or four rotations. I also like to be oriented on a cardinal heading as we enter the spin, using the section lines on the ground to give the pilot a point of reference. Once the spin begins and the nose is pointed at the ground, the section lines, tree lines, or roads are the best ground reference that the pilot has to keep his or her orientation. Maintain the ailerons in a neutral position, the stick or control yoke fully back and full rudder toward the direction of spin rotation in. If you change these control inputs, the characteristics of the spin will also change.

Before you actually begin spin training, attempt to get an idea of what the pitch angles will be during the spin from your flight instructor while you are still on the ground. I explain what pilots will see before we practice spins, but even with that knowledge some pilots are shocked by the steep pitch angles, rotational speeds, and rapid changes the plane transitions through from the stall to the spin. After a few spins, surprise is overcome and you will be able to anticipate what is going to happen, making it easier to absorb what is taking place during the spin.

We have covered planned spin entry. Now let's take some time to look at unplanned spin entry. Like unplanned stalls, the purpose of learning to do spins is to be able to recognize the potential for a spin and how to avoid it. Then, if that should fail, how to recover from it as rapidly as possible. Almost without exception, pilots who are exposed to spins for the first time have a strong tendency to pull back on the control yoke, trying to raise the nose. We have already established that you need to break the stall to stop the spin, and keeping the yoke back will prevent you from getting out. Learning what causes a spin, and how to prevent yourself from getting into one, is the best way to deal with unplanned spins.

One of the most common accidental spin situations, and the one we will focus on in this section, occurs during the landing phase of flight. According to NTSB reports I have reviewed that dealt with accidental spins, a large percentage took place during landings. Of those, some of these spins were precipitated by loss of engine power or the pilot trying to get the plane to turn tighter. One common thread in these accidents was improper coordination of controls while flying. Granted, losing the engine can put a certain amount of stress on the pilot, but you must continue to fly the airplane correctly. Most light singles found at the airport have good glide characteristics and will give you the opportunity to land safely. I know; I've been there. I lost an engine a few years ago during landing and made a dead stick landing. I made it to the airport boundaries but didn't have enough altitude to reach a runway. The plane was undamaged and the passengers walked away unharmed, none the worse for the wear. But if I had let the

Fig. 4-3A *Spin entry: forward view.*

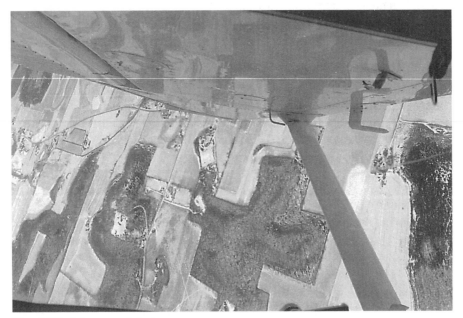

Fig. 4-3B *Spin entry: left view.*

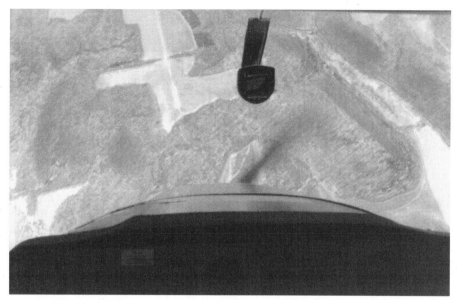

Fig. 4-4A *Forward view: in spin.*

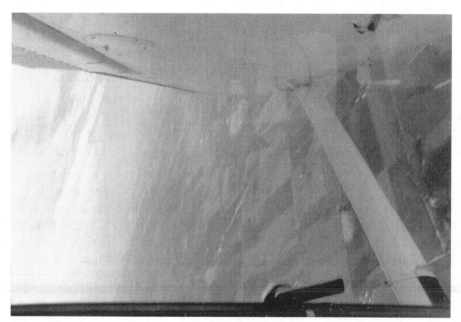

Fig. 4-4B *Side view: in spin.*

situation get ahead of me, let the airspeed get too slow, or attempted to force the plane into a turn that it couldn't complete, the outcome could have been different.

If you find yourself in an engine-out situation, don't make the banks too steep thinking this will help you get in closer to the airport or your chosen landing spot. Keep your banks within safety limits and keep the ball centered. Many pilots who try to execute very steep turns will use opposite rudder to help hold the nose up as opposed to pulling back on the elevator. As the bank steepens and the plane slows even further, this can result in a stall/spin situation.

Even if you are making a normal approach, with no other problems, making a steep turn in the pattern and not using coordinated control inputs can result in a stall/spin. Keep the ball centered as you fly, and this will help you avoid unplanned spins. One accident report I read discussed how the pilot did not like to bank the plane, instead using rudder to yaw the nose of the plane in the direction of the turn. On this particular flight, witnesses watched as the plane yawed heavily and the pilot tried to tighten up the turn to final. Unfortunately, this uncoordinated turn to final resulted in a stall/spin for the pilot.

If you should happen to accidentally enter a spin, you will need to determine which direction the plane is spinning in before you know which rudder input to make. Looking directly over the nose of the plane is the best location to look to figure out whether the plane is spinning to the left or right. Spins can be very disorienting until you get used to doing them, and an unplanned spin is even more disorienting because it catches you by surprise. Looking out the side of the plane can cause further disorientation and delay spin recovery. By looking over the nose of the plane, you can more quickly determine the spin's direction of rotation and apply the proper rudder pedal inputs. If you are so disoriented that you can't tell which way the plane is spinning, step on the rudder pedal that has more resistance. The difference in airflow over the rudder can make the rudder pedal that would be used for spin recovery more resistant to movement. And if using that rudder pedal fails, try the other pedal. Practicing spins can help reduce the level of disorientation and make it easier to determine a spin's direction of rotation. Let's move on to common errors made during spin entry.

Common Errors

The most common spin entry errors are associated with use of the flight controls as you slow the plane and enter the spin. The two most notable of these errors are insufficient backpressure on the elevator control and not keeping ailerons in the neutral position. Insufficient backpressure on the elevator control can prevent the plane from stalling cleanly and getting a good break as the stall takes place. If this happens it can make it more difficult to get the plane to enter a spin. On those occasions where I have flown with students who make this mistake, they can actually lose several hundred feet trying to get the plane to stall enough to enter a spin. We always start with enough altitude to maintain a huge margin of safety, so this does not present a problem from that aspect, but it does make it hard to tell when the plane actually transitions into the spin.

Using ailerons relates to the insufficient elevator backpressure situation. As students attempt to force the plane into the spin while it mushes along, they frequently roll aileron in, thinking their use will help things along. We already know this can cause the plane to spiral, which looks similar to the spin but is a completely different maneuver. Once the plane does spin, the use of ailerons can cause the spin to enter a flat or knife-edge mode, depending on whether the aileron input is in the same direction as the spin or opposite the direction of rotation. Unless you are flying an aerobatic aircraft rated for these maneuvers, this can result in a spin mode you may not be able to recover from or that can place excessive stress on the airframe.

Many students new to spins do not give a brisk enough input to the rudder, or they push it all the way to its stop as the plane stalls. Given that pilots are taught to execute most private and commercial maneuvers with much less than full control inputs, this is an understandable mistake. Use of less than full rudder, briskly applied as the plane stalls, may prevent the plane from cleanly entering the spin. A common reaction from the plane is that it yaws in the direction of the rudder input but never really breaks cleanly and enters the spin. You do not want to be overly forceful in the rudder input, but be authoritative in its application and you will get a nice, clean entry into the spin.

Another common error is forgetting to have the engine at idle as the plane enters the stall/spin. If power is carried into a spin, it can cause the nose of the plane to rise during the spin, putting the plane into a flat spin mode. One accident report I read discussed a pilot that added power after the spin had started, just to see what would happen. The spin turned into a flat spin, the plane's engine quit running, and the plane impacted the ground. The pilot was very lucky and survived the crash, but you should always be certain that power is at idle when you stall the airplane and plan to enter a spin. If you should happen to inadvertently carry power as you enter the spin, immediately reduce it to idle.

SPIN RECOVERY

Once in the spin, you will need to be able to recover from it. The procedures for spin recovery outlined in this section are those recommended by the FAA for general spin recovery, but you should determine if the recovery procedures for the plane you are flying differ by checking its operations manual. If the operating manual recommends a different set of procedures, use those rather than the ones documented in this section.

Spin recovery is an active process that you must initiate to cause the plane to exit from the spin. Some aircraft may recover from a spin on their own if you let go of the controls, but this is not the recommended manner to stop a spin. According to the FAA (*Flight Training Handbook*, p. 157), the proper spin recover techniques are:

1. Retard power.
2. Apply opposite rudder to slow rotation.
3. Apply positive forward elevator movement to break the stall.
4. Neutralize rudder as spinning stops.
5. Return to level flight.

Figure 4-5 shows a pilot taking these steps to recover from a spin. We have already indicated that a plane can take several turns to recover from a spin AFTER the correct

CLOSE THROTTLE
FULL OPPOSITE RUDDER
BRISK FORWARD ELEVATOR

HOLD ELEVATOR FORWARD
NEUTRALIZE RUDDER

EASE ELEVATOR BACK
TOWARD NEUTRAL

Fig. 4-5 *Spin recovery procedure.*

control inputs have been applied. Many pilots become concerned if the recovery control inputs do not immediately produce a recovery from the spin, and they attempt other inputs, prolonging the time and altitude lost before they actually stop the spin.

The rudder input opposite the rotation of the spin should be brisk and to the full limit of the rudder travel. This input stops the yawing motion of the plane and ends the rotation of the spin. The forward application of elevator reduces the angle of attack and breaks the stall. The amount of elevator input necessary and the briskness of the input will vary from plane to plane. For many aircraft, a small forward movement is enough to break the stall, while with others it will be necessary to get the elevator control forward of neutral before the angle of attack is reduced to below the critical angle of attack. Keep the ailerons neutral as you recover; it is not uncommon for pilots to roll in ailerons in the same direction as the counter spin rudder input. This can aggravate the stall, making it more difficult to recover from the spin. It can also cause the flat or knife-edge spin, as discussed previously in this section.

Once the spin is stopped and the rudder is neutral, ease the nose of the plane back to level flight. Avoid being overaggressive on the nose-up elevator input; like stall recovery, this can result in a secondary stall. At higher airspeeds, excessive use of the elevator can result in higher g-loads on the plane and damage the airframe, not to mention making for an uncomfortable recovery. In an accidental spin at low altitude, you will need to balance the need to minimize altitude loss to avoid impacting the ground with avoiding a secondary stall that could result in additional altitude loss. At best this is a very difficult recovery situation and, through spin onset recognition and spin avoidance, can be minimized.

Common Errors

Spins can be disorienting to pilots who are new to them, especially the recovery phase of the spin. They have just had the plane pitch nose down and begin to rotate as they watch the terrain below spin crazily about the nose. When it comes time to begin the recovery, the pilot is often more than ready to exit the spin, and he or she applies the recovery controls in an aggressive, and sometimes counterproductive, manner. After a little practice, most pilots begin to feel more comfortable practicing spins and settle down as they make control inputs during recovery. But during the first few spins, a number of common errors seem to come to the surface.

The first error is incorrect order of application of the controls. One pilot I flew with knew spin recovery required power-off, full opposite rudder and forward elevator control input, but he figured the order didn't matter. The order he wanted to use reversed the rudder and elevator steps, which can result in an accelerated spin. We will take a look at the accelerated spin later in the chapter, but it is not something you want to attempt on purpose in most aircraft. Since the order does matter, you need to be sure you know the spin recovery steps for the plane you fly and their order. Improper order of the steps could keep you from recovering from the spin or make the spin even worse.

The next common error is to not use enough rudder during the spin recovery. Partial rudder input opposite the direction of the spin, or slow input of the rudder, can slow the recovery from the spin. Make sure you use a brisk motion as the rudder is applied, and take the rudder pedal right to the stop. This will help ensure that you slow the rotation, a necessary step in spin recovery. Once rotation has halted, you must then remember to neutralize the rudder pedals. Many pilots will forget to center the rudder after the spin stops, and the plane begins to yaw in the opposite direction as a result of the continued rudder input. This can result in the plane entering a spin in the opposite direction of the one it just recovered from. Centering the rudder pedals can help prevent this from taking place.

Use of the elevator during spin recovery also presents a problem for pilots as they learn spins. Many pilots use an excessive amount of forward elevator as they release elevator backpressure to reduce the angle of attack and break the stall. This aggressive elevator input does two things, the first being that it forces the nose of the plane down even more steeply. This can be unsettling to a pilot who is already overwhelmed by the pitch angles the plane might achieve during a spin. The rapid, excessive forward pressure on the elevator can also cause you to float against the seat belts and cause dirt to float up from the floor. Use the minimum forward elevator needed to break the stall. Additional elevator actually increases the rate of descent and the loss of altitude. Figure 4-6 shows the altitude lost during one, two, and three-turn spins, comparing the loss between a Cessna 152 and Pitts Special S-2B. These are for comparison only and should not be considered indicative of the altitude those model aircraft will lose in a spin. Notice that a one-turn spin results in a 600- to 800-foot loss of altitude. If you are in the pattern, you don't have much room to recover if you actually get into a spin. Excessive forward elevator pressure as you recover could mean the difference between having or not having a successful recovery.

Once pilots have stopped the spin and broken the stall, they often pull back on the elevator control too rapidly or with too much force as they attempt to get back to straight and level flight. This action can result in an accelerated stall and even more altitude loss. Use the elevator with authority, but do not overreact and cause an accelerated stall during spin recovery. Not only can this result in additional altitude loss, but it can also place unnecessary g-loads on the airframe. In extreme cases, severe airframe damage can take place. At the other end of the spectrum, some students do not pull out of the dive after spin recovery quickly enough. Not only does this cause additional altitude loss, but also it allows the airspeed to rapidly build. If left unchecked, V_{ne} can be reached very quickly, again with the potential for structural damage. Learn as much as you can about the plane you fly, how it feels as it enters the spin, and what control inputs and aircraft attitudes it takes to recover with a minimum of altitude loss. Finally, remember to keep ailerons in the neutral position during spin recovery. It is not uncommon for pilots to roll ailerons

NUMBER OF TURNS	CESSNA 152	PITTS SPECIAL S-2B
1	600 ft.	800 ft.
2	1,100 ft.	1,500 ft.
3	1,500 ft.	2,000 ft.

Fig. 4-6 *Spin altitude loss table.*

in the direction of the spin recovery rudder, but like our discussion regarding ailerons and stalls, this can make the recovery from a spin more difficult.

IMPROPER RECOVERY PROCEDURE RESULTS

We have looked at spin entry and recovery so far in this chapter and to some extent discussed common errors that pilots tend to make as they learn spins. In this section we are going to look at the improper use of controls and how their misuse can affect the flight characteristics of your plane during a spin. Not every situation can be covered, and not every airplane will react the same, but the information in this section can be used as an example of how a plane might behave if proper control inputs are not made during the execution of spins.

Let's begin by looking at a very basic control, one that some pilots tend to not think of as a flight control—the throttle. Like ailerons, rudder, and the elevator, the throttle can have a very significant impact on the plane's attitude and flight characteristics. Most planes are designed so that when power is added the nose tends to rise, and when power is reduced the nose will drop if there is no intervention by the pilot. This design feature helps prevent the plane from accidentally stalling if power is lost. As power drops, the pitch angle of the nose and the plane's angle of attack are reduced. When you are in a spin, this becomes a very important consideration. If power is held during spin entry, or added after the spin begins, the nose of the plane tends to rise in comparison to the horizon. This change in attitude puts the spin into what is known as a *flat mode*. Figure 4-7 illustrates the difference in pitch angles between a normal and a flat-mode spin. Once a plane is in a flat-mode spin, it can become more difficult, and with some planes impossible, to recover from the spin. I have done flat spins many times in the Pitts Special I fly, and in the right airplane they are quite a bit of fun. But in the wrong plane they can be very dangerous, so be certain that as soon as the plane enters a spin you immediately pull the power back to idle.

The effect the rudder can have during spins is the next area we will explore. We have already seen that it is necessary to use full, brisk rudder inputs both during spin entry and recovery. Figure 4-8 shows how the combination of elevator placement and rudder can affect how much authority the rudder might have during spin recovery. The figure shows how the airflow past the rudder can be blanketed by the elevators when they are placed low in relation to the rudder. When a plane has sufficient rudder area below the elevators, it has more authority to help reduce the yaw during spin recovery. Because most general-aviation aircraft have at least part of the airflow over the rudder blanked out during a spin, it is necessary to use a brisk, full range of travel rudder inputs to improve the rudder's authority. I have had training sessions where students make a small rudder input and the plane keeps right on spinning. At that point I have the student move the rudder back to the original position, then use the brisk, full movement opposite the direction of spin rotation to stop the yawing motions. You may find that a plane seems to recover from a spin with the use of only partial rudder, but others may not. To avoid getting into the habit of not using full rudder input during recovery, always use full rudder when recovering from a spin.

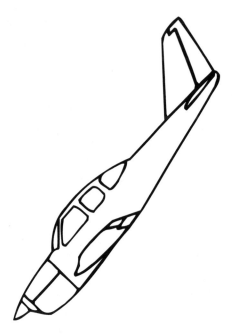

AIRCRAFT IN NORMAL MODE SPIN ATTITUDE

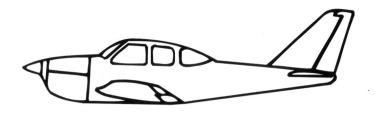

AIRCRAFT IN FLAT MODE SPIN ATTITUDE

Fig. 4-7 *Normal versus flat spin modes.*

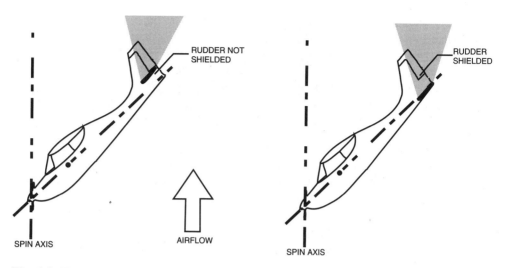

Fig. 4-8 *Horizontal stabilizer placement.*

The elevator can have a major impact on how a plane spins and how it recovers. If you do not have full aft elevator control input when you enter a spin, it may have a tendency to wallow for a longer period of time before it settles into a spin. Too much up elevator input just prior to the stall may raise the pitch of the nose to the point that the nose achieves very nose-down attitudes during spin entry. I once had a student who was practicing inverted spins push the nose of the plane extremely high just before the plane stalled. Once the inverted stall took place, the nose of the plane dropped to an extreme angle below the horizon and the rate of spin was initially quite rapid. In this particular case, the plane was designed to be able to deal with the maneuver so it was not a situation that was of concern. But some aircraft may not be designed to handle the higher g-loads and rates of rotation this spin may involve.

Once you have entered the spin, do not release backpressure until you are ready to recover from the spin. In a practice spin that involves several turns, it is not uncommon for pilots to unconsciously release some of the backpressure they are holding as the spin progresses. As long as pro-spin rudder is being held, releasing backpressure on the elevator can cause the rate of rotation of the spin to increase. This increase in the rate of rotation can be somewhat disorienting and intimidating the first time a pilot experiences it, and it can lead to further errors in the use of flight controls to recover from the spin. Make sure that once you have entered a spin, you keep the elevator control aft until you have input full rudder opposite the rotation of the spin.

We have already learned that improper use of ailerons during a stall can aggravate the stall, and the same is true regarding ailerons and spins. Previously we established that the ailerons need to remain in the neutral position while the plane is in a spin. Pilots who are in the habit of making coordinated rudder and aileron inputs often roll in aileron when they input rudder to enter a spin.

This "pro-spin" aileron can cause the plane to roll into the spin, potentially changing the spin characteristics dramatically. At the other end of the spectrum, some pilots attempt to stop a spin through "counter-spin" aileron inputs, rolling in ailerons opposite the rotation of a spin. This is actually one of the steps that is taken to purposely enter a flat spin in aerobatic aircraft. Obviously you don't want to get the average single-engine aircraft into a flat spin unless it was designed for it. As you practice spins, make sure you keep the ailerons neutral through the course of the maneuver.

EXOTIC SPINS

For the remainder of the chapter we are going to discuss some of the more exotic spins that planes can get into, but most pilots are not even aware of. For the most part you do not need to be concerned about getting training in these more unusual spins, unless you happen to enjoy the sensations regular spins produce. The purpose of this section is to make you aware that these types of spins exist and give you more detail regarding some of the spins that were mentioned earlier, including inverted and flat spins. Under the right circumstances and in the right airplane, the spins we will cover next are a great deal of fun.

There is a certain amount of myth surrounding these spins, especially the flat spin. We have all heard stories about pilots who got their airplanes into flat spins and the only way out was to drop the flaps, or the landing gear, and somehow the plane suddenly responded. Movies like *Top Gun* and *Firefox* seem to encourage a fear of flat spins, and spins in general. I've done hundreds of spins, upright and inverted, in aerobatic planes, and they are absolutely safe as long as the plane is capable of executing the maneuver and you know what you are doing. Get a qualified aerobatic flight instructor to go up with you if you decide you want to pursue practicing these spins. If you do go for the instruction, make sure the plane is qualified and that you are wearing parachutes as you practice. The regulations state that if the plane achieves bank angles in excess of 60 degrees, or pitch angles greater than 30 degrees, a parachute is required. Once you meet these criteria, be prepared to have some serious fun.

Inverted Spins

Pilots who fly aerobatics spend a fair amount of time in the inverted position, with good reason. It's a lot of fun. There's definitely a thrill when you roll the airplane upside down, look up, and see the ground. Part of any aerobatic training should include inverted spins for a number of reasons. Many advanced routines require flying what are known as outside maneuvers where the pilot is pulling negative g's. We know slow flight in the positive part of the envelope can result in normal spins, and slow flight from the inverted position can result in inverted spins. Knowing how inverted spins feel can help the pilot recognize them, just as with upright spins, and provide for more rapid recovery. And since accidental spins seem to be part of learning aerobatics, this is a great way to avoid any unnecessary surprises.

Figure 4-9 shows an external view of a plane in the inverted position, in a slow flight mode with a high angle of attack. Just as with normal spins, the pilot reduces

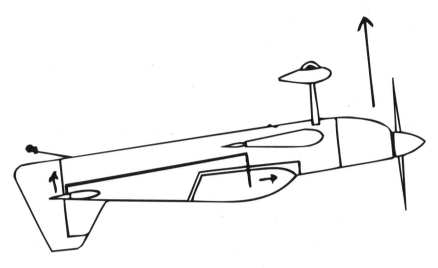

Fig. 4-9 *Inverted slow flight attitude.*

power and slows the plane by increasing the pitch of the nose while maintaining altitude. The difference between inverted and normal stalls is that the elevator control is moved forward to raise the nose of the plane when it is inverted, instead of aft. As the plane slows, the pilot can feel buffet set in and the nose will break downward. As the plane stalls in the inverted position, the pilot now pushes the control stick fully forward to hold the plane in a stall. As the same time the pilot gives a left or right rudder input using a brisk, full range of motion to cause the plane to go into a spin. Figure 4-10 shows the plane now in a relatively steep nose-down pitch attitude. The plane will go through the spin entry phase, the incipient spin phase, developed spin phase, and spin recovery phase. Most airplanes have a tendency to have a higher rate of rotation during inverted spins as compared to upright spins. The g-forces also tend to throw you away from the plane, which can be disconcerting to pilots who are new to inverted spins.

As the plane continues in the spin, the pilot must maintain full forward stick to keep the plane stalled and full pro-spin rudder. I have heard some misconceptions that some pilots have voiced regarding rudder input and spin direction. Some students I have flown with believe that the plane will spin the opposite direction of the rudder that is input. This is not true. If you input full left rudder as the spin entry input, the plane will spin to the pilot's left; if you put in right rudder, the plane will spin to the right from your point of view.

Being pilots who are addicted to inverted spins, we will let the plane go through several rotations as the spin progresses. Like upright spins, the nose of the plane will rise and fall through the course of the spin, and the rate of rotation will change during the spin. We must be certain that we keep the ailerons centered to help avoid the flat spins and knife-edge spins we discussed in upright spins.

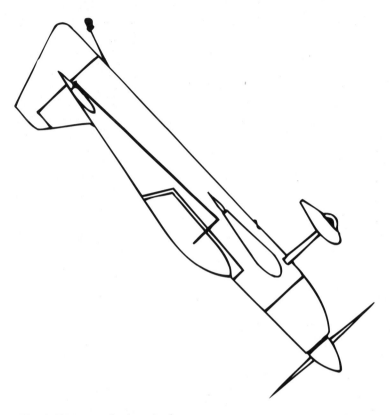

Fig. 4-10 *Inverted spin attitude.*

Recovery procedures from inverted spins are very similar to upright spins; the steps include:

1. Power off.
2. Full rudder opposite the spin's direction of rotation.
3. Release forward pressure on the elevator control, reducing the angle of attack to below the critical angle of attack.
4. Neutralize the rudder input.
5. Raise the nose to level flight.

You can see there is little difference between flying inverted spins and upright spins, except that inverted spins are a whole lot of fun. One point that should be made is that as you raise the nose to level flight, you will be pushing forward on the control stick. Since you are upside down, this gets the nose back to a level flight attitude. As you practice inverted spins you will no doubt notice immediately the steeper pitch angles and more rapid rates of rotation as compared to upright spins. Different airplanes also have different inverted spin characteristics. My experience has been that a Decathlon has a shallower pitch angle and rate of rotation as compared to a Pitts Special S-2B, which tends to spin more rapidly and with the nose further below the horizon.

Common Errors

Many of the errors pilots tend to make during upright spins also are present during inverted spins. Pilots sometimes have trouble getting a good, clean break as the plane stalls in the inverted attitude. This is very common during initial training in inverted spins due to the extremely new flight attitudes and use of controls. As the plane stalls, make sure you get the stick forward in a smooth motion to assure that the critical angle of attack is obtained throughout the spin's entry. Another common error is waiting to input rudder in the desired direction of the spin, or using insufficient rudder to start the yawing motion. Again, this can be attributed to the new attitudes and sensations a pilot experiences while learning inverted spins. The plane can wallow if the controls are not used properly, ending up in an inverted dive or spiral instead of a spin.

While in the inverted spin, it is not uncommon for pilots to relax the forward pressure on the control stick. This tendency is even greater than that of pilots to release backpressure during upright spins. In order to get full forward stick in many planes it is necessary to actually lean forward. As the inverted spin progresses, pilots tend to relax back into the seat and the stick moves aft as a result. Releasing the forward pressure can result in an accelerated spin, making the rate of rotation increase. Since inverted spins already have a higher rate of rotation, this can be a shock to pilots the first time they experience one. Maintain full rudder into the spin as it progresses, keeping in mind that pilots will also tend to relax their use of rudder pedals since they are concentrating on keeping their feet from falling away from the floor of the aircraft. Inverted flight is a great deal of fun, but the new physical, mental, and visual experiences can be overwhelming during initial inverted experiences.

When exiting the inverted spin, make sure you use full opposite rudder to stop the yaw. Given the higher rate of rotation, it is more likely the plane will take longer to recover or not be able to recover if full rudder is not used. After full rudder opposite the direction of rotation is applied, release the forward pressure on the control stick and break the stall. The nose of the plane will very likely be lower than compared to an upright spin, so altitude loss can be higher during inverted spin recovery. Try to get the nose back to level flight as soon as possible without pushing forward too quickly. This can result in the potential for a secondary stall/spin. The increased negative g-forces can also be more uncomfortable than the positive g's you experience, and excessive forward pressure can cause additional discomfort.

Accelerated Spins

We have noted throughout our discussions of spins that you want to keep the stick fully aft during upright spins and fully forward during inverted spins. The term accelerated spin has been used, but now let's take a look at what an accelerated spin actually is and why it should be avoided in most aircraft. An accelerated spin takes place when a plane is in a spin and the angle of attack is reduced before the rotation caused by yaw is halted. In an upright spin this takes place when the backpressure on the elevator control is reduced; in an inverted spin the relaxation of forward elevator input causes an accelerated spin.

I have executed quite a few accelerated spins, both upright and inverted, while giving spin training in a Pitts Special. For most pilots, the realization of just what an accelerated spin feels like is somewhat intimidating the first time they experience them. With just minor forward elevator inputs during an upright spin, the rate of rotation rapidly increases. When full forward stick is applied, the rate of rotation becomes quite impressive. This increase in spin rotation can be disorienting and more physically demanding on the pilot as the g-forces increase along with the rate of rotation. Most pilots express a certain amount of surprise in how fast a plane can rotate when it enters an accelerated spin. These additional physical forces also affect the structure of the aircraft, with the rotational and g-force loadings possibly overstressing some aircraft, even though they may be rated for spins.

I have researched accelerated spins but have been unable to document what causes the plane to accelerate as the angle of attack is reduced with relaxed elevator backpressure. Figure 4-11 illustrates my theory on what causes this increase in rotation rate. In a normal spin the inside wing is stalled, while the outside wing is generating more lift than its counterpart. This is depicted in the upper portion of the figure. As you relax elevator backpressure, the angle of attack is reduced, allowing the outer wing to reduce the amount of area that is stalled and resulting in an increase in the amount of lift produced. The inside wing is still past the critical angle of attack and the lift differential between the two wings increases, the outer wing now generating even more lift than the inner wing. As a result of this increased lift differential, the rate of rotation increases significantly.

I remember the first time I did an accelerated spin. I had done hundreds of normal spins, and even though the instructor explained what I should expect, I was unconcerned. After all, this was just another spin and I was in a Pitts Special so this should be fun. I entered a normal spin and let it settle into a nice spin, then pushed the stick forward. I was amazed at how the plane reacted and how the g-forces increased with the greatly increased rotation. Knowing I was in a Pitts, which is very sturdy, made me feel comfortable with the situation, but I was glad we had lots of altitude for this spin session. After several turns I initiated spin recovery procedures and the little airplane recovered without a problem. In fact, we did several more accelerated spins because it was quite a bit of fun and I wanted to keep doing them.

On the ground I reflected on what an accelerated spin would be like for a pilot uninitiated in spins, and in particular, accelerated spins. I can imagine that it would be quite easy to make the wrong control inputs, which would make the situation even worse. Later in this chapter we will cover crossover spins, which takes an upright spin and turns it into an inverted spin. To make that transition, the plane must enter an upright accelerated spin, and control inputs are made that cause it to go inverted in the blink of an eye. We will wait until the cross-over spin section before we get into those details, but realize that accelerated spins pose a special situation that requires you to use the correct procedures to recover from them.

RAISED WING HAS NOT
STALLED AND IS
PRODUCING GREATER
LIFT THAN STALLED WING

LOWER WING
HAS PASSED CRITICAL
ANGLE OF ATTACK AND
IS PRODUCING LITTLE LIFT

NORMAL SPIN LIFT DIFFERENTIAL

RAISED WING IS NOW
GENERATING EVEN
GREATER LIFT THAN
STALLED WING

ACCELERATED SPIN LIFT DIFFERENTIAL

Fig. 4-11 *Accelerated spin lift differential.*

Recovery procedures from an accelerated spin do not use the standard spin recovery procedures we discussed earlier. To review, the FAA procedures for normal spin recovery are:

1. Retard power.
2. Apply opposite rudder to slow rotation.
3. Apply positive forward elevator movement to break the stall.
4. Neutralize rudder as spinning stops.
5. Return to level flight.

When you are in an accelerated spin, it can be difficult to get the controls in a position that will change the spin mode from an accelerated mode to a normal spin mode. In the Pitts Special, the Muller-Beggs spin recovery procedures are used. We will discuss those procedures in more detail later.

There are several points you should remember regarding accelerated spins. First, you do not want to enter one unless the plane you are flying is capable of executing them safely. Aerobatic aircraft like the Pitts Special are a good example of planes that can safely execute an accelerated spin. Before you ever practice accelerated spins, verify that the airplane is capable of executing and recovering from one. The second point to consider is that if you use proper spin recovery control inputs, in the proper sequence, the likelihood of ever entering an accelerated spin is greatly reduced. Finally, once in an accelerated spin, normal spin recovery inputs may not be effective. The success of normal spin recovery inputs in recovering from an accelerated spin will depend on the conditions and the plane. If the plane does not respond, the emergency spin recovery procedures we will discuss may be helpful, but if the plane is not designed for these spins they may not work successfully.

We have covered what can cause an airplane to enter an accelerated spin, a theory about what causes the increased rate of rotation, the complications of practicing accelerated spins, and that it may be necessary to use different spin recovery procedures to recover from accelerated spins. One thing we have not discussed is how the average pilot gets into an accelerated spin accidentally.

The most likely scenario for accidental entry into an accelerated spin can be found in the comments of a pilot I mentioned earlier in the chapter. He knew the basic spin recovery procedures but felt that the sequence of control inputs was not important. If pilots with this understanding should enter an accidental spin, they would remember several steps:

- Power off.
- Full opposite rudder.
- Forward elevator control.

If our imaginary pilot executed the forward elevator control before stopping the rotation of the spin with full opposite rudder, the plane would enter an accelerated spin. Our pilot, now concerned that the plane is not recovering from the spin, might then input full opposite rudder, knowing that is yet another step that must be taken. At this point the pilot has set up for the spin to turn into a crossover spin, and the plane could enter an inverted spin. I have spoken to aerobatic pilots who have accidentally gotten into just this situation, and it happens so fast they do not realize the plane went inverted for several rotations.

This example should make it clear that correct control inputs in the correct order can avoid accelerated spins. If you are interested in getting flight instruction in accelerated spins, find a qualified flight instructor using an airplane capable of advanced spin training. I highly recommend this type of training for pilots who intend to do spins on a regular basis, and it is almost mandatory for aerobatic pilots to get this type of training. I enjoy throwing myself around the sky, and understanding these advanced spins makes for a safer pilot.

Crossover Spins

Crossover spins are a combination of several spins we have discussed so far. Not generally well known, they are most common during aerobatic flight when a pilot misinterprets a spin situation and makes the wrong control inputs. In this section we are going to explain how pilots get into crossover spins and how to avoid them. Unless you are flying aerobatics, the likelihood that you will ever encounter a crossover spin is very small. Knowing about them, what control inputs can cause them, and how to avoid them are the areas we will concentrate on.

We have already covered normal spins, accelerated spins, and inverted spins. These are components of a crossover spin, steps that a pilot must pass through to get from an upright spin to an inverted spin. Let's take a look at the steps. When you start a normal upright spin, you will have the elevator control fully aft, and pro-spin rudder. This is normally held until you are ready to recover from the spin. You already know that if you apply spin recovery steps in the wrong order, this can aggravate the spin, so let's assume you push the forward stick before you input the correct rudder for spin recovery. The forward stick will put you into an accelerated spin, but you are in an upright accelerated spin. Next, you decide you must need to input rudder opposite the direction of the spin. This action will cause the plane to flip into an inverted spin, almost in the blink of an eye. I have executed crossover spins where the transition was so smooth it was hard to tell the plane went into an inverted spin, and others where I was thrown against the seat belts and knew exactly when the plane transitioned.

The problem with crossover spins is that many pilots do not know what has happened or what steps to take next. They have already input what they think are spin recovery controls, and the plane keeps spinning! In reality, after the plane transitions from the upright spin to the inverted spin, the control inputs actually keep the plane in an inverted spin. You'll recall from our inverted spin discussion that forward stick is necessary to maintain the angle of attack needed for a stall in the inverted spin. The rudder input also happens to be pro-spin rudder, so the plane continues on in the inverted spin.

Recognizing that you have transitioned from an upright spin to an inverted spin is the first step in recovery from the crossover spin. I attended an aerobatic safety seminar several years ago during which one of the speakers related his experience with crossover spins. He was flying an airshow, and part of his routine consisted of a ten-turn inverted spin. During his recovery from the spin he accidentally let the plane get into an accelerated, inverted spin, then pushed opposite rudder, not realizing what was taking place. The plane went from an inverted spin to an upright spin so smoothly he did not realize what had happened until about three more rotations took place. When he finally figured it out, the pilot applied the correct control inputs and the plane recovered.

Hopefully you will never be placed in this position. The purpose of this section is to let you know that incorrect spin recovery procedures can make things more exciting than you planned, and it is important to know the correct spin recovery procedures for the plane you fly. Should you find yourself in a spin that normal recovery procedures do not work for, try the Muller-Beggs emergency spin recovery technique that we will discuss in the next section of this chapter. It does not work for all aircraft, but it offers another opportunity to recover from a spin.

EMERGENCY SPIN RECOVERY

In addition to the normal, FAA-approved spin recovery procedures we have discussed, there is another recovery procedure. First publicized by Mr. Eric Muller and now promoted by Mr. Gene Beggs, this procedure is based on findings that planes are able to recover from spins if a simplified spin recovery procedure is used. This procedure reduces the pilot's need to correctly position the elevator and ailerons for spin recovery, or to need to know what type of spin they are in. To summarize, the procedure is:

1. Cut that throttle!
2. Take your hands off the stick!
3. Kick full rudder opposite until the spin stops!
4. Neutralize rudder and pull out of the dive! (*Sport Aerobatics*, p. 31, April 1994)

You can see that this procedure differs from the recommended version by the FAA. One of the largest differences is releasing the control stick, which allows the airflow to position the elevator and ailerons in spin recovery positions. Once rudder opposite the direction of the spin is applied, the plane recovers from the spin.

In the Pitts Special I fly, I have used the Muller-Beggs spin recovery procedure hundreds of times during recovery from normal spins, inverted spins, upright and inverted flat spins, and accelerated spins. I teach this method of spin recovery to my students because it simplifies recovery from some of the more exotic spins we have discussed. Especially in aerobatics, it may take time to figure out the type of spin you may have accidentally entered, and this decreases the time for spin recovery.

Mr. Beggs has successfully tested the recovery procedure in several aircraft with success, as long as the plane was loaded within proper weight and balance ranges. The aircraft tested by Mr. Beggs included several Pitts Special models, the Christen Eagle II, the Cessna 150, Cessna 172, and the Beechcraft Skipper trainer (*Sport Aerobatics*, p. 31, April 1994).

I must mention certain cautions regarding this spin recovery technique, though. NASA studies have found that the procedure does not work with all aircraft and should not be depended on in all cases. Some authors have suggested that the Muller/Beggs recovery procedure may work at certain points in a spin while not working at others.

Spin rotation rates, angle of attack, or other factors may influence how well this spin recovery procedure works in a particular plane. If you are not certain it will work in the plane you fly, contact the manufacturer. As always, the manufacturer is the final word in the correct spin recovery procedures for a given plane.

FLAT SPINS

The last topic we will cover in this chapter is flat spins, one of my favorite spins to execute. Often misunderstood, they are a great deal of fun if flown in the right airplane and are as safe as any normal spin. In this section we will review what a flat spin is, what can cause an airplane to enter a flat spin, and how to get out of flat spins. A word of caution before we go any further, though. Flat spins are the most entertaining spins to fly, in my opinion, but they can be extremely dangerous if flown in the wrong airplane or without the proper flight training. DO NOT attempt flat spins on your own without first learning the proper methods for entering and recovering from a flat spin in a plane capable of executing them safely. Remember, safety should always be the highest priority on any flight.

Flat spins have a mystery about them that has been passed on from pilot to pilot until the words "flat spin" conjure up an airplane that is completely out of control without any chance for a controlled recovery. We've all heard stories about the pilot who accidentally got his planes into a flat spin and managed to get out by lowering the landing gear or some other technique that mysteriously got the nose of the plane to drop and the spin stopped. Or the stories about the Luscomb or Champ pilot who got into a flat spin and could not get out, only to walk away from the crash scene because the impact was soft enough that no one was seriously injured. While these make great hangar flying tales, there is a great deal of information about flat spins. Unfortunately, most of it is passed along among aerobatic and airshow pilots and does not make it to the typical general-aviation pilot.

Let's begin our discussion with what causes a plane to enter a flat spin. There are actually a number of ways that planes might enter a flat spin, in fact more than we can cover in this section, so we will discuss some of the more common methods. In all cases the plane must first enter a normal spin, which transitions to a flat spin. The transition might be very quick, but you must first get the airplane into a spin, whether upright or inverted. For pilots who are trying to purposely transition to a flat spin, a number of steps are necessary. Let's take an example of a pilot who wants to get the plane into an upright flat spin with a spin direction to the left. Like any normal spin entry, the pilot stalls the plane with the power off, then pushes full left rudder with the stick in the fully aft position. As we would expect, the plane will enter a normal, upright spin to the left and continue to spin in that manner if no changes are made and the airplane is within the center of gravity envelope.

Once the plane is in the spin, the pilot does several things to change the spin mode from normal to flat. First, while maintaining full left rudder, the pilot pushes the ailerons fully right. This has the effect of rolling the airplane to the right and brings the wings more parallel to the horizon. Next the pilot adds engine power, which raises the nose and puts the longitudinal axis more parallel to the horizon. This reduced nose-down angle, in combination with the wings-level attitude resulting from the right aileron, causes the plane to flatten its attitude and change from a nose-down spin mode to a flat spin mode.

Sitting inside a plane that is in an upright flat spin is very much like sitting on top of a Frisbee. The plane seems to be in a normal, flat attitude but it is spinning around its center of gravity in a very flat manner. I really enjoy watching passengers and students the first time they experience a flat spin. They are invariably awed by what the plane is doing and the sensations they feel as the plane behaves in a very unusual manner. In fact, many people comment that a flat spin is actually more docile than a normal spin. The plane spins slower in the flat mode as compared to a normal spin and really gives quite a nice ride as it continues in the flat spin. Since it is difficult or impossible to see the ground very well in a flat spin, and even tougher to pick out a landmark to judge the number of rotations, I tend to use the sun as a reference point for keeping track of how many turns we have completed.

To recover from a flat spin, you must transition the plane back from the flat mode to a normal mode, and recover. I use and teach the Muller-Beggs emergency spin recovery technique to all the pilots I train. I have used it hundreds of times in recovering Pitts Specials from flat spins, and it works extremely well. I learned it from the instructor who taught me how to fly flat spins, and I have continued to use it. As we discussed, the steps for that recovery are:

1. Cut that throttle!

2. Take your hands off the stick!

3. Kick full rudder opposite until the spin stops!

4. Neutralize rudder and pull out of the dive! (*Sport Aerobatics*, p. 31, April 1994)

I know of pilots who state that they don't trust the plane to recover on its own, but this does work in the Pitts Specials I have flown. The plane you learn flat spins in may have different techniques, so follow the steps recommended by the manufacturer. With the Muller-Beggs technique, cutting the throttle helps get the nose down, reducing the flat spin tendency. Letting go of the control stick lets the ailerons center, and the wings tend to return to a normal spin mode, the inside wing dropping. You already know that full opposite rudder stops the yawing or rotation of the spin. Releasing the elevator will tend to let the nose drop and reduce the angle of attack, stopping the stall. Once again, this technique may not work for all planes, but if you have tried everything else and you are out of options, it may come in handy to know this spin recovery technique.

We won't get into a lot of detail on inverted flat spins, but they are very much like upright flat spins. Entry is achieved by entering an inverted spin, adding power, then using aileron in the same direction of the rudder. Because you are inverted, this has the effect of leveling the wings as in the upright flat spin. Inverted flat spins are also a lot of fun to fly. In fact, most airshow pilots like to use inverted flat spins in their routines as opposed to upright flat spins because less smoke gets into the cockpit as the plane spins down. Recovery is achieved through the Muller-Beggs technique.

We've talked about using controls to enter a flat spin, but what else can cause a spin to flatten out? Extended flaps can cause some planes to flatten their spin mode, while others are less affected by having the flaps out. If you do enter a stall/spin, it might be a good idea to retract the flaps to avoid this tendency. Keeping power on after the plane enters a spin can also cause the plane's nose to rise, and this alone might be enough for some planes to enter a flat spin. Getting power off as soon as a plane starts to spin can also help avoid accidental flat spins. Finally, aft center of gravity can cause a plane to flatten out its rotation. Figure 4-12 shows how centrifugal force can cause the weight loaded aft of the center of gravity to begin to lift the tail in relation to the horizon, causing the spin mode to become flat. This is why it is very important to observe center of gravity restrictions when practicing spins. Once the plane gets into this type of flat spin, it may be impossible to recover from it. At least one crash of an aerobatic airplane has

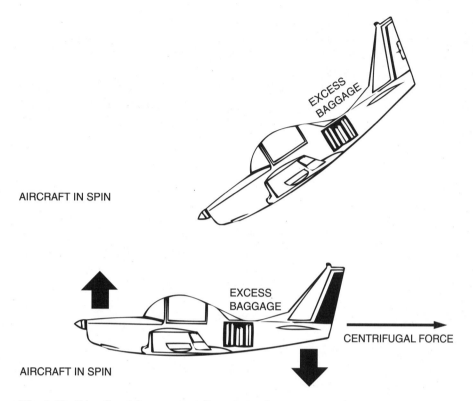

Fig. 4-12 *Centrifugal force creates flat spin mode.*

been attributed directly to the pilot loading the plane incorrectly and then doing aerobatics that ended up in a flat spin and fatal crash.

This section has briefly covered flat spins, not so much to teach you how to fly them, but how to avoid them. If you make sure you chop the power when you start a spin, use the controls correctly, and keep the plane loaded within its allowed center of gravity, you can reduce the chances of accidentally entering a flat spin. If you're interested in learning more about flat spins, I highly recommend finding a qualified training program to get some hands-on experience. It is absolutely one of the most enjoyable and exciting experiences you can undertake in flying unusual attitudes in an airplane.

CONCLUSION

This chapter covered a great deal of material in a relatively short span. I have written an entire book on stalls and spins, which gives significantly more detail and photographs that can help you understand what takes place in each of the spins discussed in this chapter. By understanding spins, what causes them, and the mostly likely situations that can result in an accidental stall/spin, you can increase your chances of avoiding these dangerous scenarios. It is extremely important that you use the flight controls in a coordinated manner, are vigilant in watching your airspeed, and understand how bank angles affect stall speed. If you should happen to enter an accidental spin, knowing how to react to recover quickly will reduce the amount of altitude you lose, which is very important if you are at pattern altitudes.

The discussions regarding the exotic spins—inverted, accelerated, and flat—are more to make you aware that these types of spins exist and how to avoid entering them through improper control use while in a normal spin. Having experience in each of these spins is very important for aerobatic pilots because they have a higher probability for entering them, but realizing how improper control use can cause a normal spin to turn into one of these spins is important to the average pilot.

In closing this chapter, I'd like to stress that spin training is one of the most important pieces of training any pilot can get. Many pilots don't understand spins, and don't want to. There are many reasons, all of them invalid. I am a strong believer that many of the accidents each year that are the result of an accidental stall spin could have been avoided if the pilot had received even one or two hours of spin training from a qualified flight instructor. Challenge yourself, make yourself a safer pilot, and go get spin training. You, your family, and your passengers will never regret that you did.

5
Ground Reference Maneuvers

WE HAVE COVERED SOME VERY IMPORTANT TOPICS SO FAR IN THE book: basic aerodynamics, basic flight maneuvers, stalls and slow flight, and spins. Each of these topics has centered on how to fly the airplane through some basic, but extremely important, maneuvers. As we go forward in the book we are going to begin to learn about those areas that are necessary for you to pass various flight tests for the private and commercial licenses. The topics so far have taught us how to fly a plane in straight and level, turns, climbs and descents, and slow flight. Now we will continue to build on the basic flight maneuvers.

Ground reference maneuvers are designed to teach us how to compensate for the effect wind can have on our track across the ground. We will learn why knowing how to correctly compensate for the wind's effect is important and what types of problems we can run into if we do not take the correct wind correction steps. As you take off and land, it is important that you be able to maintain your distance relative to the runway, preventing yourself from drifting toward or away from the runway. Each of the maneuvers described in this chapter are designed to help you learn to anticipate and correct for the wind's effects.

We will review a number of ground reference maneuvers, including S-turns across a road, turns about a point, and rectangular course. Eights around pylons and eights on pylons will also be reviewed. How these maneuvers apply to landing patterns will be discussed to help you understand the relevancy of these exercises in compensating for the effects of wind as you fly in the pattern.

None of the ground reference maneuvers we will discuss in this chapter are particularly difficult to master. For this reason I think quite a number of pilots tend to forget about the importance of maintaining a level of competency and understanding for each of the ground reference maneuvers we will cover. I have flown with too many pilots who do not crab the airplane into the wind correctly during takeoff and landing, the result being that the pattern they fly is not parallel to the runway, nor is the distance constant. Take the time to think about each maneuver as you read this chapter, and then go practice it. You will find that it will make you more aware of how the wind affects you while you fly.

PURPOSE OF GROUND REFERENCE MANEUVERS

During the introduction to this chapter, I hinted at why we need to learn and remain competent at flying ground reference maneuvers. As a private pilot, you were required to learn S-turns across a road, rectangular course, and turns about a point. If you have been through commercial pilot training, you also learned how to fly eights around pylons and eights on pylons. Each of these maneuvers taught you a different aspect of correcting for the effects of wind on your ground track. When you first learned to fly these maneuvers, you probably had to consciously think about the direction the wind was blowing from in relation to the direction the plane was flying, but after practicing a little it became easier to anticipate how the wind was going to affect the plane in relation to the reference you were flying around.

After learning to fly the ground reference maneuvers with a reasonable degree of proficiency, you were then able to correct for the effects of the wind while flying in the pattern. If you were like me, you probably let the airplane drift away from or toward the runway while you were flying the downwind leg of the pattern. I got the plane parallel to the runway, but the wind kept pushing us from the ground track I wanted to fly. After learning how to fly ground reference maneuvers, I began to think about the relationship of the wind, the direction the plane was flying, and what I needed to do to compensate for the wind—crabbing into or away from the wind, making turns that are greater than or less than 90 degrees to make sure the ground track was a 90-degree turn. Learning to change my angle of back depending on the wind's effect on my ground speed also helped me to avoid flying past the point I wanted to hit, such as turning from base to final. And how many times as students did we all watch the centerline of the runway drift off to one side or another as we flew down final approach to the runway?

With an instructor watching us, we began to fly square patterns, maintaining a constant distance from the runway. We were better able to keep the ground track of the plane aligned with the centerline of the runway, and we managed to make a turn from base to final without flying through the final approach course. That ability to judge how to compensate for the wind came directly from mastering ground reference maneuvers.

The problem is that many of us forget to keep practicing those skills, and they begin to deteriorate. I have flown with quite a number of pilots who let the airplane's ground track go where the wind pushes them. This is most evident when they are turning from base to final and consistently end up flying past the runway centerline, or even if they get things lined up initially, they drift off to the side as they fail to compensate for the wind. While this may not seem like much and you might tell yourself there's always time to correct, pilots make mistakes that cause accidents as they suddenly realize they are overshooting a turn, or they have been blown in too close to the runway while on downwind and they attempt to compensate by using very steep banked turns as they transition from downwind to base or base to final. These situations are where the stall/spin accidents we discussed in Chapters 3 and 4 arise, and this is the reason we need to remain proficient in compensating for the effects of wind on our ground track as we fly. In this chapter we are going to learn how wind affects our ground speed and ground track. We are also going to learn how to crab into the wind and change our angle of bank during turns to maintain a constant ground track. Let's begin with a relatively easy maneuver, S-turns across a road.

S-TURNS ACROSS A ROAD

S-turns across a road are a great maneuver to help you learn to anticipate how wind affects your ground speed and turn radius. By using a road as a reference point, you are able to see just how far the wind can blow you off your desired course as you fly a series of 180-degree turns across it. In this section we are going to learn what an S-turn across a road is and the effects that wind can have on the ground course we are attempting to follow. We will also learn how to compensate for the effects of wind by changing our angle of bank throughout the entire maneuver. I like teaching S-turns to students and pilots who are interested in a refresher because it is a simple-looking maneuver to fly, but it demonstrates the need to know where the wind is from and how to correct for it so graphically.

Figure 5-1 is an illustration of an S-turn across a road. You can see that the plane transcribes a series of S's across the road as you make a number of consecutive 180-degree turns, first in one direction, then the other. Let's break the maneuver down into four phases: entry into it, a turn to the left, a turn to the right, and exit from the maneuver. In each phase we will review how the controls should be used, what the effect of the wind is, and the proper corrections you will need to make to compensate for the wind.

Figure 5-1 indicates that the best conditions for practicing S-turns is a road that is perpendicular to the wind. While this is often impossible, you should attempt to find a road that is closest to fitting that relationship. You should also find a road that is relatively long; you will want to be able to fly a number of turns back and forth across it. It can become frustrating if you must stop after just a few turns because you run out of road and are unable to judge whether you are maintaining the correct ground track. You will want to maintain approximately pattern altitudes during the course of the maneuver, and the altitude should remain constant throughout the entire series of turns.

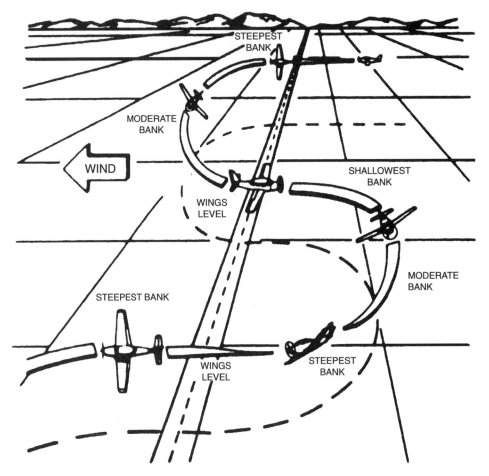

Fig. 5-1 *S-Turn across a road.*

While flying the S-turn, you want your ground track to maintain a constant radius through each 180-degree turn. Because wind direction will affect your ground speed, you will change your angle of bank and roll rate throughout the maneuver to compensate for the wind's effects. When your ground speeds are the highest, you will need to use the greatest angle of bank, shallowing the bank as the turn carries the plane into a crosswind, then upwind heading. The shallowest portion of the bank will be when the plane's heading is nearly into the wind. The figure illustrates this fact, showing that the steepest bank is necessary just after you cross the road on a downwind heading, becoming a moderate bank as the plane reaches a crosswind heading, and becoming a very shallow bank just prior to the plane crossing the road on an upwind heading. Another point to consider is the fact that you will constantly be changing the plane's crab into the wind throughout the turn. When you are flying in a direction parallel to the wind, you will have no crab included, and when you are flying perpendicular to the wind, you will have the greatest amount of crab angle.

There are several aspects that you must consider as you fly the S-turn maneuver. Approach the road from the upwind side on a downwind heading. I like to use pattern altitudes for this maneuver, somewhere between 700 and 1000 feet AGL. When you are flying in a downwind direction, you will have the highest ground speed, which will require the greatest amount of bank just after you cross the road. As you continue the turn, monitor the radius and shallow the bank slowly to keep a constant distance from the imaginary center of the semicircle you are flying. I like to have students use distances similar to the distance they keep the plane from the runway on the downwind leg of flying the pattern, which can range anywhere from one-quarter to one-half mile, depending on the plane, speeds, and conditions. If you attempt to fly a radius that is too tight, you will need to use excessively steep banked turns during some portions of the maneuver and will find it difficult to cross the road perpendicular to it. If the distance is too far away, you will have a difficult time judging whether the radius is constant, and the bank will become extremely shallow during the upwind portion of the turn.

Begin the initial bank immediately after crossing the road on the downwind heading, the steepest banked portion of the turn. Experience will teach you how steep to make the initial bank, and this will depend on wind speed, aircraft speed, and the length of the radius for the ground track. As you reach the 90-degree point of turn, you should be at a moderately steep bank; continue to shallow the bank further as you turn to a more upwind heading. The wings should be level and the direction of flight perpendicular to the road just as you cross the road. You will then begin a turn in the opposite direction of the downwind turn. Since the plane is now at an upwind heading, you will begin with the shallowest bank, increasing it slightly as you turn towards a crosswind heading and continuing to increase the bank angle as you reach a downwind direction of flight. The steepest part of this turn will be just prior to rolling the wings level as you cross the road. Once again, as you cross the road you will begin yet another turn in the direction opposite from the one you just completed and continue the maneuver. At this point you will begin the turn with the steepest bank you will need and continue to shallow it as you change directions relative to the wind.

Pilots tend to have trouble judging the correct angle of bank through various parts of an S-turn. This becomes easier as the pilot gains more experience with different wind conditions. There is also a tendency to gain or lose altitude throughout the maneuver as a result of constantly changing the angle of bank. This is a very good maneuver to help pilots learn to keep their eyes outside of the cockpit and learn to fly by looking at ground track and the feel of the airplane. Many of the situational awareness issues that we discussed earlier in the book now begin to come into play. I like the S-turn maneuver because it lets the pilots see how well they are doing at compensating for the wind and teaches them how to anticipate the effects wind will have on their ground track.

TURNS ABOUT A POINT

Turns about a point are another good maneuver designed to teach pilots how to compensate for the effects of wind on their ground track. As the name suggests, you pick a point, usually a road intersection, and fly a series of circles around it with the center of the road acting as the center of the circle. You will normally fly at least two complete circles around the ground reference point while maintaining a constant radius from it. Figure 5-2 shows a turn about a point. Like the S-turn, you want to maintain a constant altitude during the maneuver, usually in the 700 to 1000 foot range, and use banks no greater than 45 degrees at the steepest point in the turn. The closer you are to the point you are flying around, the steeper the bank that will be necessary at the downwind point of the turn where the ground speed is greatest.

Turns about a point use changing bank angles and roll rates to compensate for the changes in ground speed that take place due to the changing effects of wind as you change directions during the turn. The steepest bank will be necessary when you are on a direct downwind heading, and the shallowest bank will be used when you are headed into the wind. Moderate bank angles will be used when you are at crosswind headings during the turn. Like the S-turn maneuver, your crab angles will be greatest when you are perpendicular to the direction of the wind and at zero when you are heading directly into or away from the wind. The fact that you are changing bank angles and heading make the fact that you are actually crabbing the airplane into the wind more difficult to realize, but the crab in necessary to maintain the constant turn radius you need.

As you begin to set up for the turn about a point maneuver, find a good landmark to use as the center of the turn. A road intersection or tree that is easily visible from the air will work well, but you want to be sure you are not flying near communities, farms, livestock, or homes while executing this maneuver to avoid annoying those on the ground. Set up to enter the turn on a downwind heading at a distance that you want to be the radius of the turn. You will immediately enter the bank that will be the steepest part of the turn, as illustrated in Figure 5-2. As you change directions you will begin to shallow the bank and increase the crab you are using while you turn to a crosswind heading. The crab

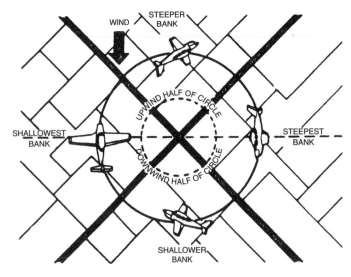

Fig. 5-2 *Turns about a point.*

shown in the figure is exaggerated to illustrate what is taking place during the turn. The angle of bank continues to become shallower as the plane approaches an upwind heading, with the wings at their shallowest point when your heading is directly upwind. As you continue the turn, you will once again be increasing the angle of bank as the ground speed of the plane increases until you are downwind again and using the steepest bank needed.

Like S-turns, it can be difficult to judge the correct angle of bank as you enter the turn. When pilots first learn ground reference maneuvers, they often chase after the correct turn radius, at first being blown too far away, then using very steep bank angles to get back to the correct point. With practice, pilots can anticipate how much bank they will need as they enter the turn and then are able to judge how well they are doing at maintaining a constant radius. The length of the radius is also important. If you are too far away in a high-wing airplane, the lowered wind may block the pilot's view of the center of the turn. As you become more skilled at turns about a point, you will be able to judge the correct radius and bank angle as you enter the turn. This skill is important as you fly the pattern and need to maintain a constant altitude and distance from the runway as you progress from downwind to base to final. Common errors include gaining or losing altitude during the maneuver and not being able to maintain a constant radius. Excessive bank angles are also common as pilots learn to judge what effect wind is having on their ground speed and turn radius.

RECTANGULAR COURSE

The rectangular course maneuver applies directly to flying an airport traffic pattern. Figure 5-3 depicts an airplane flying a rectangular course. The maneuver is accomplished by picking out a rectangular area, a farm field works very well, and flying at an equal distance from the boundaries of the field. Once again, you will need to compensate for the wind's effect throughout the maneuver through the use of crabbing into the wind and varying bank angles. Like S-turns across a road and turns about a point, you want to maintain a constant altitude as you fly the rectangular course. We will discuss the crab and bank angles needed during each leg of the maneuver in this section. You can enter a rectangular course from any direction, but for the purposes of this section we will assume entry is as seen in the illustration, on a downwind heading.

Use a distance outside of the boundaries of the rectangular course of approximately one-quarter to one-half mile. This provides sufficient distance in most cases to keep a view of the edges of the field throughout the maneuver, whether you are flying using left or right turns. In the case of this discussion, we will use turns to the left, but the concepts apply for both directions. During the downwind leg of the rectangular course in this example, you will have no crab since the wind is directly at your back. In the real world, this is rarely the case, and you will likely need some crab throughout the maneuver. As you come abeam the corner for the first turn, the intersection of roads in this case, you will begin your turn. Since our ground speed is highest at this point, you will need the steepest bank required during the maneuver. As you continue through the turn, you will want to maintain a constant radius from the corner, which is the center point of turn. As

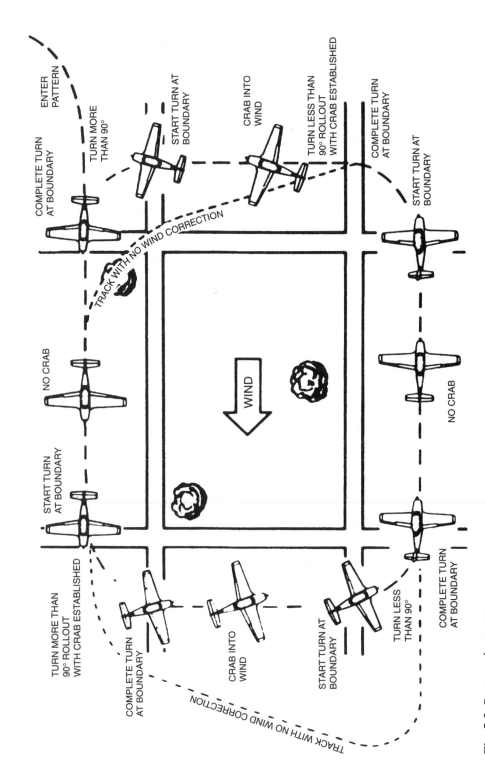

Fig. 5-3 *Rectangular course.*

we have already learned, we will need to decrease the bank as the ground speed slows. On this particular turn we will have a crosswind from the left, which will have a tendency to push us away from the ground course we want to maintain a constant distance from. In order to compensate for this effect from the wind, we will actually turn more than 90 degrees so that when we roll out of the turn, we are crabbed into the wind. The figure illustrates this crab into the wind after the turn is completed. The amount of crab will depend on the speed of the wind, your airspeed, and the actual direction the wind is coming from.

As the maneuver continues, we will continue to fly along the second leg of the course until we are once again abeam the turning reference point, in this case the second road intersection. We will want to maintain a constant radius through the turn. In this case we are starting out crabbed into the direction of the turn and will end up turning less than 90 degrees when the turn is completed. The bank angle will be moderate at the start of the turn and be reduced as we turn into the wind. In this example we are flying directly into the wind after the turn is completed and no crab will be necessary. In real life we will most likely need to crab to some degree in each leg of the rectangular course. At this point we continue to fly the third leg of the course, maintaining a constant distance from the road we are using as a reference point. The distance should be equal to the previous two legs of the course. Since we are flying into the wind, our ground speed will be the slowest of the maneuver.

The third turn will start as we come abeam the next turn reference point, the road intersection. Because our ground speed is slow, we will start out with a very shallow bank and increase it to a moderate bank before finally rolling out. The wind is now blowing from our right, pushing us back toward the road; as a result we will want to turn less than 90 degrees to set the plane up in a crab into the wind to allow us to maintain a constant distance from the road. This crab is also illustrated in the figure, and it should be noted that the crab should be established as part of the roll to wings level from the bank, not established after the wings have been rolled to level. We will continue flying parallel to the road, monitoring the distance we are flying from it. If necessary we can increase or reduce the amount of crab angle to hold the proper distance from the road we are using as a reference point. Once again we continue to fly parallel to the road until we are abeam of the next turning reference point.

The fourth turn in the rectangular course will be greater than 90 degrees in our example, since we are crabbed into the wind. We will start our turn with a moderate bank, increasing it as we turn onto a downwind heading due to our increasing ground speed. The steepest portion of the bank will be just prior to beginning the rollout to wings level. Since the wind is directly at our back, we will establish a heading that is parallel to the road. We have now covered each turn and leg of the rectangular course; the course can be repeated as many times as necessary to become familiar with the bank and crab angles necessary to maintain constant distances from our ground reference.

It is not uncommon for pilots to lose altitude during turns, or to let the plane drift toward or away from the ground reference point. Until pilots become accomplished at judging the effects wind will have on their ground track, it takes some adjusting to establish the correct crab angles and bank angles. With little practice, it is possible to correctly anticipate how much correction is needed, to quickly judge if the correct amount has been used, and readjust to maintain parallel ground tracks. I like this maneuver from the standpoint that it teaches pilots how to put together changing bank angles during turns to maintain a constant radius from the reference point, and how to crab to fly a course parallel to a given ground reference. Pilots learn to quickly recognize drift due to wind and correct for it, which is a needed skill when flying in the traffic pattern.

ADVANCED GROUND REFERENCE MANEUVERS

Ground reference maneuvers for the commercial pilot become somewhat more complicated than those we have discussed so far. The private pilot is expected to understand what the effects of winds are and how to correct for them in the pattern and during cross-country flying, but the commercial pilot is held to a higher standard. In this section we will cover those advanced ground reference maneuvers, eights around pylons, and eights on pylons. Both of these maneuvers are more demanding of the pilot's ability to fly by reference to objects on the ground, compensate for the effects of wind, and control the airplane in an acceptable manner. Some of the added complexity is that the plane now flies figure eights around two "pylons," each pylon being the center point of one-half of the figure eight. Crabbing and bank control become very important in successfully flying each of these maneuvers.

Eights around Pylons

Figure 5-4 illustrates an eight around pylons. As you can see, two reference points, in this case trees, are used as the center of the circle for each half of the figure eight. This maneuver gives the pilot the chance to perform wind correction through turns to both the left and right while maintaining a constant altitude. The two points should be between

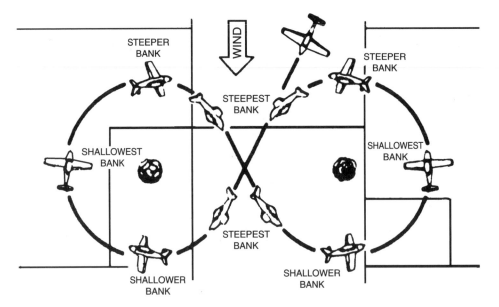

Fig. 5-4 *Eights around pylons.*

one-half and one mile apart, and this distance can be adjusted depending on the speed of the plane you are flying and the speed of the wind. As with other ground reference maneuvers, you should perform eights around pylons away from homes, farms, livestock, etc., to avoid annoying those on the ground.

The line between the pylons you choose should be perpendicular to the direction of the wind. Fly the maneuver at pattern altitudes and enter between the two ground reference points on a downwind heading. In our example you will be making your initial turn to the right. Like turns about a point, you want to maintain a constant distance from the pylon you are flying around, but now you will need to transition smoothly from flying one circle, then another, each in opposite directions. Since we are entering on a downwind heading, the bank will initially be steep, then become more moderate as we turn to a crosswind heading. It will shallow further at the upwind heading portion of the maneuver, then increase in bank as we turn to crosswind, then downwind once again. Crab, bank, and roll rate will apply in this maneuver just as they did in S-turns and turns about a point, helping you maintain a constant distance from the point while you compensate for the effects of wind on your ground track.

The amount of bank needed as you enter an eight around pylons will depend on your distance from the pylon, the speed of the wind, and your ground speed. You already know that the bank will shallow until you have reached an upwind heading, then will steepen again as you approach a downwind heading. As you complete the first circuit of the pylon, you will then transition to the second portion of the maneuver. In this case the turn will now be to the left. As before, the turn will initially be the steepest as you enter the turn to the left, then shallow to the least amount of bank at upwind, then steepen again as your heading becomes more downwind.

Eights around pylons help you learn to judge your distance from an object on the ground, compensate for the effects of wind, and give you experience in dealing with the more complex task of judging wind and how it affects your ground track as you move between two points. It is not uncommon for pilots to focus so much on one aspect of the maneuver that they ignore other portions of it. Some may have a tendency to lose altitude, or to let the wind blow them from the desired course, but with a little practice the maneuver is not difficult to master.

Eights on Pylons

While an eights on pylon generates a figure eight around two pylons like the eights around pylons, it is a significantly different maneuver. The eights on pylons maneuver differs from the other ground reference maneuvers we have discussed in this chapter in several respects. First, it is not necessary to maintain a constant radius from the pylons we are flying the maneuver around and second, the altitude will not remain constant during the course of the maneuver. The real purpose of flying eights on pylons is not so much to use it as a ground reference maneuver, but to help you learn to divide your attention between flying the airplane and looking outside the cockpit.

Figure 5-5 illustrates what an eights on pylons maneuver looks like as you are flying through it. You will notice that there are three factors that are mentioned at various

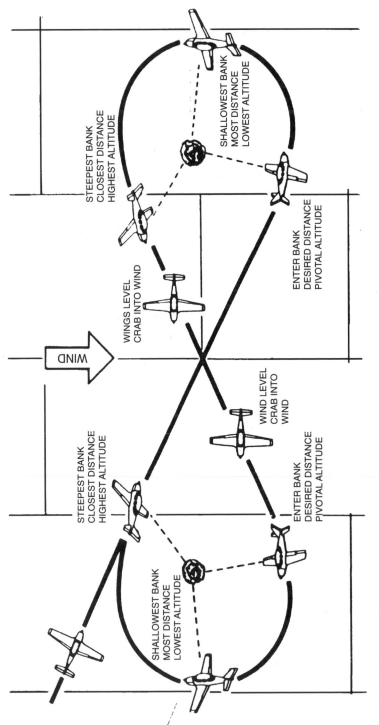

Fig. 5-5 *Eights on pylons.*

portions of the maneuver: altitude, bank angle, and distance from the pylons. We already know that the distance will change, and it is understandable that the bank angle will change, but no doubt you are wondering about the altitude changes. Let's discuss why the altitude changes and how we determine when we are flying the maneuver correctly.

During the course of an eights on pylons, you will fly the airplane at an altitude and airspeed that result in a line parallel to the airplane's lateral axis, extending from the pilot's eye, appears to pivot on each of the pylons you fly around. The altitude you fly at to accomplish the maneuver, known as the pivotal altitude, is determined by your ground speed. This line of sight parallel to your eye is occasionally compared to the position of the wing tip, but only under certain circumstances will this be the case. Depending on your seat position and the placement of the wing, you may find that the correct line of sight is ahead of or behind the wing tip; you will need to experiment in a particular plane to determine exactly where you want the pylon to hold during the maneuver. Figure 5-6 shows the difference in the line of sight the occupant of an airplane would see when viewing a pylon from the front and rear seats. As you can see, the pylon has moved to a position behind the wing when you move to the rear seat of the plane. Figure 5-7 shows how using the correct versus the incorrect reference point will affect whether the correct radius, or arc, is flown around the pylon. Picking the incorrect reference point can make the turn radius too small, causing the plane to move in too close to the pylon, or too far away from it.

The pivotal altitude will change with the ground speed of an airplane; the angle of bank will not affect the pivotal altitude. Now we begin to see why we will change our altitude as our position relative to the wind changes, and therefore our ground speed changes. Referring back to Figure 5-5, you will notice that when we are headed into the wind we have our shallowest bank, our greatest distance from the pylon, and the lowest altitude. As our speed increases we increase the bank angle, move in closer to the pylon, and increase our altitude. Obviously, the faster our ground speed, the higher the pivotal altitude, and the lower the ground speed, the lower the pivotal altitude. Figure 5-8

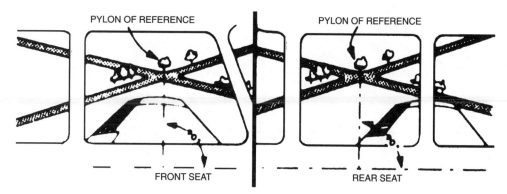

Fig. 5-6 *Eights on pylons: line of sight.*

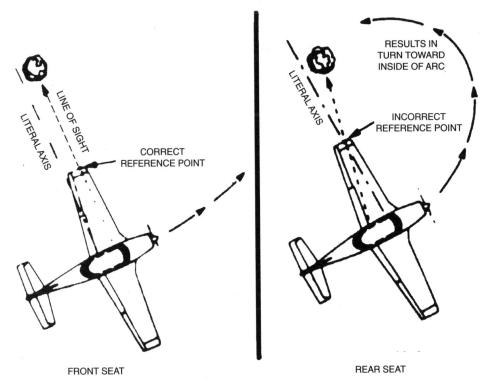

Fig. 5-7 *Eights on pylons: reference points.*

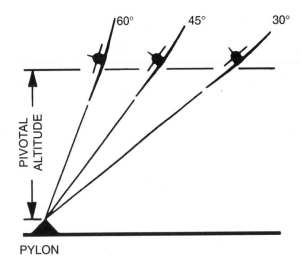

Fig. 5-8 *Bank angle versus pylon distances.*

documents that the steeper the bank angle is, the closer in you will be to the pylon, and the shallower the bank, the further away from it you will be. You can begin to see how this maneuver becomes particularly busy for pilots, as they must take into account the visual position of the pylon, the direction the wind is from, the ground speed, bank angle, and altitude of the plane on a continuous basis! I like it because it teaches you to fly the airplane based on what you see, hear, and feel as you watch outside the airplane while you make all of these changes.

Now that we have covered some of the mechanics of the maneuver, we will walk through actually flying it. You want to pick two pylons that are easy to see and are separated enough to allow 3 to 5 seconds of straight and level flight between them as you roll out of one bank and before you roll into the next bank. The pylons should be perpendicular to the direction of the wind, as well. You will enter on a downwind heading diagonally between the pylons, crabbing into the wind while the wings are straight and level. You will start the turn when the pylon appears to be just ahead of the wingtip, lowering the wing and placing your line of vision on the pylon. Maintain your line of sight reference by gradually increasing the bank; the reference line should appear to pivot on the pylon.

As the turn continues, the plane will begin to fly an upwind heading, the ground speed slowing as a result. Due to the reduced ground speed, we will need to lose altitude, descending during the turn to maintain the correct pivotal altitude. You will accomplish this by easing forward on the elevator control just enough to hold the pylon in your line of sight reference as the turn progresses. When you begin to enter the crosswind portion of the turn, the wind will push the plane closer to the pylon and you will need to increase the angle of bank to hold the proper reference point. As we fly further into a downwind heading, the bank will become steeper, the ground speed will increase, and the altitude will need to increase. To accomplish the altitude gain, you will need to ease back on the control yoke and cause the plane to climb just enough to once again hold the reference point on the pylon. As you enter the diagonal portion of the maneuver, you will roll the wing level, crab into the wind, and wait until the pylon reaches the point just ahead of the wingtip to begin the next portion of the turn. The same technique for holding the correct reference line on the pylon applies to the second half of the maneuver as the first.

During the turn, if the reference point begins to move ahead of the pylon, you should increase altitude; if the reference line moves aft, the pilot should decrease altitude. A way to visualize this is that you will move the elevator control AWAY from the position the pylon is moving toward to maintain the correct altitude. How much control movement is necessary will depend on the airplane's characteristics, but you will be able to anticipate the correct control inputs with a little practice. Be sure NOT to use the rudder to force the nose of the plane to move in an attempt to put the reference point on the pylon; this results in uncoordinated use of the controls and the problems associated with it, which we have already discussed. In fact, improper use of the rudders to control the relation between the reference point and the pylons is a very common mistake that pilots who are learning eights on pylons are prone to make.

CONCLUSION

We have discussed the basic ground reference maneuvers you should learn as you progress as a pilot. Don't just learn them, then forget them after the flight test is completed; there are some very basic concepts that each of the maneuvers discussed in this chapter teaches you. Being able to judge the effects of wind on your track over the ground is very important to flying good, solid, controlled patterns. Too many pilots just let the wind blow them where it will and attempt to "catch up" by banking too steeply or flying past the correct final approach course, then flying back toward it again.

If you should happen to lose an engine, you need to be able to judge exactly where you are going to end up and be able to adjust for the wind to allow you to land in the safest area within gliding distance. Try engine-out practice on a windy day and see if you come up short, or long, for the field you are using as your practice landing area. Many pilots can't predict whether they will really land in the field they have chosen until it is too late to make any corrections that will be useful.

There are many other ground reference maneuvers in addition to those discussed in this chapter. Find a qualified flight instructor and go practice as many different types as you can find in as many different wind conditions as possible. I like the challenge of staying ahead of the wind and airplane and keeping the airplane where I want it to be, not where the wind happens to push me.

6
Takeoffs and Landings

Takeoffs and landings are a critical phase of every flight. We are at low altitudes, at slow airspeeds, and must contend with the effects of weather, radio traffic, and other aircraft. But our passengers often judge our piloting skills by how well we do during takeoffs, and especially landings. We have all had flights with others on board where we get that perfect squeak of the tires as we touch down on the runway with a feather's lightness and hear the passengers comment on what a smooth landing that was. As pilots we even judge each other based on how well we land, how close we are to the desired touchdown point, and how smooth the touchdown is.

In this chapter we are going to look at how to become proficient at takeoffs and landings, so that we can become consistent and safe in how we execute each of these maneuvers. The subject of takeoffs and landings has filled books, so we will not be able to review it to the depth we might in a book dedicated to them, but we can begin to learn what we need to know to achieve smooth, safe takeoffs and landings consistently. We will look at what factors are important during takeoffs and landings, the control inputs, how planes typically react during each phase of a takeoff or landing, the effects of wind, and a number of other areas. We will also look at how to fly the traffic pattern, how to control your airspeed during landings, and how to pick a particular touchdown point and land on it. This chapter will focus mainly on tricycle-gear aircraft, though much of what we will cover applies to taildraggers as well.

Chapter Six

As you read the chapter, try to visualize what we are discussing and apply it to take-off and landing situations you have encountered. I really enjoy landings, trying to pick a spot on the runway during downwind and touching down on it, or as close as I can get—no matter what the conditions. As I was learning how to fly I had an instructor tell me he could evaluate pilots just by having them fly around the pattern once. The use of controls at slow airspeeds, coordinated control inputs, wind correction in the pattern, airspeed control, and ability to judge the landing zone can all be observed through one circuit of the pattern. You can also see how well pilots do at dividing their attention between the airplane, other traffic, the radio, and anything else that comes along. It should be very apparent to you why each of the preceding chapters are so important and are building blocks to becoming a pilot that safely, consistently flies the airplane through the pattern and touches down on the runway as intended. Let's begin the chapter with some basic ter-minology that we need to understand.

V-SPEEDS

V-speeds, or the various critical airspeeds you need to know about the airplane you fly, are an important part of takeoffs and landings. Stall speeds should be uppermost in your mind as you take off and land—our discussions of accidental stalls/spins being an example of why airspeeds are important. V-speeds provide a standard way to define what each of the speeds important to an airplane are. Let's define a few V-speeds you should be concerned with.

Stall Speeds

You will fly your plane in different configurations, and these configurations affect the stall speed of the airplane. To differentiate between the different airspeeds at which a plane can stall, V-speeds have been developed to designate the configuration of the plane and its stall speed. Depending on the plane, there may be a number of different V-speeds associated with different aircraft configurations, from flaps up, to partial flap extension, to full flap extension, gear extended, gear retracted, and anything else that might be important to the pilot. But for the purposes of this book, we are interested in two V-speeds related to stalls, V_{so}, or the stall speed in the landing configuration, and V_s, the stall speed when the gear and flaps are retracted. Let's begin with V_{so}.

V_{so} (Stall Speed Landing Configuration)

V_{so} is the airspeed your airplane will stall at in the landing configuration. This configura-tion can vary from plane to plane, but generally is gear down, flaps down, and power off at maximum landing weight. Pilots will often refer to this configuration as "dirty" because of the additional drag caused by having the gear and flaps extended into the airstream. V_{so} is important because you want to maintain an airspeed that is approxi-mately 1.4 times V_{so} on the downwind leg of the pattern, with the airspeed reduced to 1.3 times V_{so} as you reach short final. ("On Landings, Part I," p. 3)

This allows you to avoid becoming excessively slow while in the pattern, yet avoids airspeeds that are too high at touchdown and can result in the plane floating down the runway until you are slow enough to land. If winds of any speed are present, especially crosswinds, you may need to fly at a higher airspeed to provide enough control authority to maintain control and compensate for the effects of the wind. If the operations manual for the plane you fly provides different airspeeds for use during landings, use those as opposed to the ones recommended by the FAA guidelines.

V_s (Stall Speed Clean Configuration)

The other stall speed we will review is V_s, or the stall speed of the plane in the clean configuration. This means the plane has the flaps (and landing gear, if the plane is a retractable) retracted. Planes normally stall at a higher airspeed in the clean configuration due to the loss of lift resulting from the flaps being retracted.

Keep in mind that V_{so} and V_s stall speeds are for when the plane is in level flight. Figure 6-1 shows a table that documents how the stall speed of an airplane increases its bank angle. As you can see, in a 60-degree bank the stall speed increases approximately 1.4 times the speed it stalls at in level flight. This means an airplane that has a V_s of 44 knots will now stall at 62 knots in a 60-degree bank. It's easy to see how low, slow turns with a steep bank angle can cause you to stall the plane.

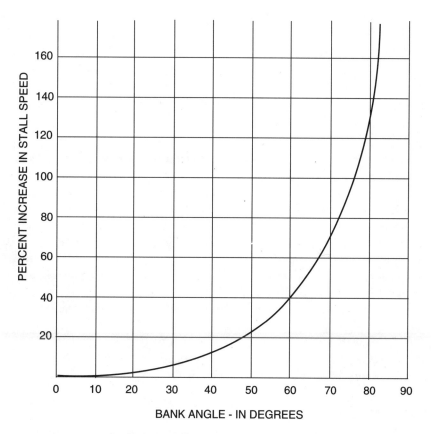

Fig. 6-1 *Bank angle versus stall speeds.*

V_{fe} (Flap Extension Speed)

V_{fe} refers to the maximum airspeed at which flaps should be in the extended position. If you fly the airplane at speeds that are greater than V_{fe} with the flaps extended, it is possible to cause physical damage to aircraft structures. The flap connecting rods, rollers, tracks, or the flaps themselves could be bent or ripped from the plane if you exceed V_{fe} with the flaps extended. The same damage could take place if you extend the flaps above V_{fe}. The white arc on the airspeed indicator shows you the safe range of airspeeds for the flaps.

V_{ne} (Velocity Never Exceed)

V_{ne}, or the airspeed above which the airplane should never be flown, is a very important number to keep in mind. Indicated by the red line on the airspeed indicator, V_{ne} is the maximum speed an airplane can safely be flown at in smooth air. V_{ne} is determined by the aircraft manufacturer based on engineering design and aircraft testing. If you should happen to exceed V_{ne}, you can cause severe damage to aircraft structures up to the point of total failure. This damage can manifest itself as structures that bend, or in serious cases, *control flutter*. Control flutter is when the control surfaces of the airplane begin to vibrate uncontrollably. These vibrations can cause damage to connecting points for the control surfaces. In severe cases of control flutter, not only can the control surface be torn from the plane, but also vibrations can become so severe that the structures they are attached to can be ripped from the aircraft. If you are flying at V_{ne} in rough air, the loads placed on the structures of the airplane as a result of the turbulence can also cause damage to the aircraft. The one thing you should remember is: Never exceed V_{ne}. You could find yourself in a very serious situation.

V_{a} (Maneuvering Speed)

Maneuvering speed, known as V_{a}, is the airspeed at or below which rapid movements of the controls or turbulence will not cause damage to the aircraft at gross weight. V_{a} is not normally marked on the airspeed indicator but may be placarded on the instrument panel. If it is not placarded, you should look it up in the operations manual for the plane you fly. If you should encounter turbulence, you should reduce the airspeed you are flying at to at or below V_{a}. If you fly at these airspeeds, the plane will stall before it generates loads that are sufficient to damage the aircraft. Of course, flying into very severe turbulence can cause damage no matter what airspeeds you are flying at, but under normal circumstances, flying below V_{a} will prevent damage.

At speeds above V_{a}, turbulence or rapid control movements may generate g-loads that exceed the structural design of the plane and may damage the plane to the point of structural failure. Review the operations manual for your aircraft; it is very likely that V_{a} values will change as the gross weight of the airplane changes. Be certain to use the correct V_{a} value for the weight of the plane you are operating.

V_{x} (Best Angle of Climb Speed)

The best angle of climb speed for an airplane allows you to gain the greatest amount of altitude within a given distance. Best angle of climb is the airspeed often used by aircraft manufacturers for climbs after a short-field takeoff or when there are obstacles you must clear at the end of the runway.

Due to the relatively high angle of attack that is achieved during a climb at V_x, the engine may have a tendency to overheat; for this reason you should avoid prolonged climbs at best angle of climb. Airflow through the air inlets is reduced as a result of the high angle of climb; if used for an extended period of time, you may find the engine runs at temperatures above recommended ranges. After you have cleared the obstacle, you should reduce the angle of attack to V_y or the airspeed recommended by the manufacturer for extended climbs.

V_y (Best Rate of Climb)

The best rate of climb will allow the plane to gain the largest amount of altitude over a given period of time. V_y is often recommended by aircraft manufacturers for extended climbs to cruise altitudes. The angle of attack used during a climb at V_y airspeeds is lower than V_x, allowing better airflow through the engine compartment and lower engine operating temperatures. You will notice substantial differences in engine oil temperatures during the summer between flying at V_x or V_y. To help save wear on the engine, and increase engine life, use V_y during extended climbs.

TRAFFIC PATTERNS

In this section we are going to take a look at what the traffic pattern is, how to enter the pattern from a number of different points, and how to exit it. The traffic pattern can be a high-traffic environment, and maintaining an awareness of other aircraft and their relation to your plane can demand a great deal from the pilot. Airplanes fly at different airspeeds, and unfortunately, at different altitudes while in the pattern. You must be able to maintain proper separation from those planes, adjusting your flight path and airspeed to remain a safe distance from traffic. Whenever you fly, you should be looking for traffic, but this becomes extremely important when you fly in the traffic pattern. The maneuvers we have covered so far in the book become of fundamental importance due to the need for dividing your attention between looking outside the plane, monitoring your ground track, looking for other airplanes, and flying the approach. Knowing how to enter and exit the pattern can help you be more efficient as you fly near the airport.

Preparation for Pattern Entry

You should begin planning for your landing while you are still well away from the airport. In a controlled airport, you will need to listen to the approach and tower frequencies, normally at distances of 20 miles or greater, depending on the airspace around the airport. At uncontrolled fields you should begin listening to the Unicom frequency while you are at least 10 to 20 miles from the airport. However, in either case, you should begin to build a mental picture of the traffic, the runways in use, the flow of traffic pattern, and any other factors that can affect how you will fly in the pattern.

Building this mental picture of what is going on around the airport can help you begin to plan how you will enter the pattern, the most likely areas to look for other aircraft, and what the types of planes in the pattern are. Knowing these facts can help you adjust your airspeed, know where you might have problems with planes that are flying significantly faster or slower than you, and what distance you may need to fly from the runway. By planning ahead, you can increase the level of safety as you fly into the pattern. As you approach the airport, you should clearly state your intentions, altitude, and position from the airport. This is true whether you are flying into a controlled or uncontrolled field.

At controlled fields, the tower staff will direct you through the airspace, giving you headings and altitudes. You are not relieved of the need to look for other aircraft, though, so don't become complacent and assume you do not need to keep looking outside the plane. At uncontrolled fields you may want to overfly the airport before you actually enter the pattern to get an idea of the runway layout, look at the windsock to determine wind direction, or observe for other aircraft in the pattern. There are times when pilots do not use their radio, or the plane they are flying is not equipped with a radio, so even though you may not hear any other traffic over the radio, make sure you are looking for other aircraft as you fly over the airport or prepare to enter the pattern. I have had quite a number of situations where airplanes that did not report their entrance or progress in the pattern at various airports flew along in front of me. As I enter the pattern, I really give the airspace a look in all directions, including behind me, to make sure there are no surprises.

Whether you are being directed by the tower at a controlled field or are flying the pattern at an uncontrolled airport, make sure you check the wind sock and compare the information you are getting from it with what the tower or Unicom staff is telling you. No one will purposely give you wrong information when you ask for weather information, but you want to use the wind sock to confirm the wind direction as you fly through the pattern. On some airports the wind sock may be difficult to find, but do your best to locate it prior to landing. Figure 6-2A shows the typical wind sock, while Figures 6-2B and 6-2C show two other wind indicators that may be at airports. Both the wind tetrahedron and landing tee are designed to point into the wind, or in the same direction you should land the plane. At some airports they may also be lighted, making it easier to locate them at night. Any of these devices can give you an idea of the actual direction of the wind near the ground and runway. This can be important if you are making any type of a crosswind landing for judging the amount of crab or slip you are going to need during the approach and landing. The airport is full of information that can help you learn and adjust your pattern as you approach the airport and fly through it; use the radio and other items like the wind sock to adapt and fly safely through the pattern.

Pattern Entry

In this section we are going to look at several different positions you might use to enter the pattern. These entry points are likely areas that can help you integrate with other traffic as smoothly as possible, providing you with good visibility and the opportunity to scan for other aircraft. Depending on the airport you are flying at, and the type of pattern

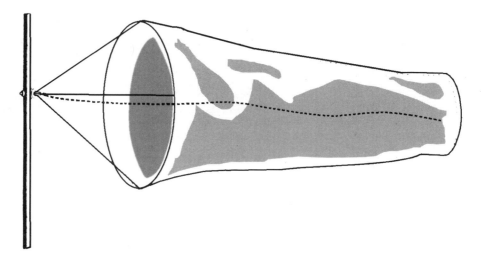

Fig. 6-2A *Wind sock.*

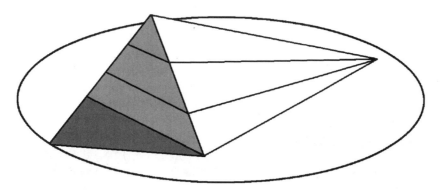

Fig. 6-2B *Wind tetrahedron.*

in use, you may find that some of these entry points do not work well. If that is the case, adapt and use an entry point that provides separation from other aircraft and a good view of the pattern from the cockpit so that you can keep an eye on other traffic.

Most airports have traffic patterns that are 800 to 1500 feet AGL in altitude. You will need to consult a sectional chart or other reference to find out what the pattern altitude is at the airports you plan to fly into. Using the correct pattern altitude is important for several reasons. First, if you are flying at the correct altitude it gives you a reference altitude to look for other traffic at. I like to scan above and below the pattern altitude as I enter and fly through the pattern, as well, because not everyone knows or uses the correct altitude. Second, it also puts you in the best position for other traffic to see you. As you fly each leg of the pattern, you should report your position so that other airplanes know where you are. Being at the correct altitude for the pattern will make it easier for pilots to find you and maintain proper separation.

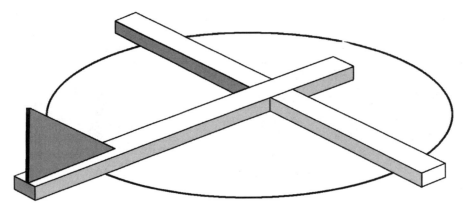

Fig. 6-2C *Landing tee.*

The Pitts Special I fly tends to need to remain high on final due to its very steep descent once power on the engine is pulled. For this reason at uncontrolled fields I like to let other traffic know that I will remain high on final so that they know where to look for me, and what my intentions are. On one occasion another plane turned onto the runway I was setting up to land on when I was on short final; the other plane sat there, then took off, causing me to go around because the pilot did not see me even though I had stated my intentions on the radio. It's a good idea to look and listen as you fly to make sure you know what traffic is doing and where it is at in the pattern.

Finally, before we move onto the legs of the pattern and the entry points, you should be aware that while normal patterns at uncontrolled airports are flown using left turns, known as a left-hand pattern, there are some airports that use right-hand patterns. Residential areas, ground obstructions, and towers in the vicinity of a left-hand pattern may cause the airport to set up a right-hand pattern. You will need to consult airport reference guides, available at most Fixed Based Operators (FBOs), to find out if an airport has right or left traffic patterns. Depending on the layout, some airports have right-hand patterns for one runway and left hand for another, so if you are flying into a field you are unfamiliar with, check with Flight Service or consult the airport reference guide before you take off.

The entry points we will cover in this section are downwind entry, crosswind entry, base leg entry, and the straight-in approach to final. Figure 6-3 shows the different legs of a pattern; you can see that they consist of the crosswind leg, the downwind leg, the base leg, and final approach. The figure also shows two runways and how the pattern is left hand in some cases, while right hand in others. I like this particular illustration because it makes pilots aware of the fact that while most uncontrolled fields use left-hand traffic patterns, there are airports that have traffic patterns that do not fit the norm. Also note the wind cone, which shows the direction that airplanes should land as a result of the wind direction.

The legs of the pattern are based on the direction of the wind. The crosswind leg is flown perpendicular to the active runway. Point 5 on the figure shows an airplane in the crosswind position. The downwind leg is flown going with the direction of the wind, parallel to the runway. The illustration also shows that with the plane in position 2 on the downwind section. Realize that if the wind is not blowing directly down the runway, there will be a crosswind that will affect your ground track as you fly through the pattern. Now is when the ground reference maneuvers, which allow us to become proficient at adjusting our flight path based on the wind to control our ground track, come into play. Our goal is to fly the ground track depicted in the figure regardless of the crosswinds that may be present in each leg of the pattern. As with the ground reference maneuvers we have previously discussed, you will want to maintain a distance of approximately one-half mile from the runway for the downwind leg.

The discussion we just had covers what goes on at an uncontrolled field. When you are flying at a controlled airport, tower personnel will direct you regarding the headings and altitudes they want you to fly. They will normally use a standard pattern consisting

THE TRAFFIC PATTERN

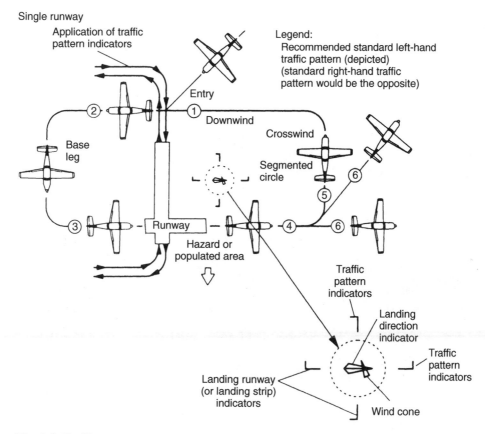

Fig. 6-3 *Traffic pattern.*

of a downwind, base, and final, but traffic considerations at the airport may cause them to issue instructions that have you fly an abbreviated pattern, or in some cases, an extended pattern. It is not uncommon for the tower to issue the instruction for a pilot to be extended downwind and the tower will call base. This is normally done for spacing between aircraft, but is an example of how flying at a controlled airport can change the normal pattern. Now let's move on to discussing pattern entry options.

Crosswind Entry

Figure 6-4 shows several pattern entry points. Position A in that illustration demonstrates a crosswind pattern entry. This is normally used when you are approaching the airport from a direction opposite the side of the runway the pattern is being flown on, and provides an efficient method of entering the flow of traffic. This pattern entry is accomplished by flying across the active runway at pattern altitude perpendicular to the runway. At the appropriate distance, you will then turn to the downwind leg of the pattern. The position of the crosswind leg should be such that aircraft taking off from the runway will not climb into you and you have the ability to maintain visual contact with departing traffic. This position is normally near the end of the active runway but may vary depending on conditions. You should be careful when making a crosswind pattern entry for other reasons, however. If traffic volume in the pattern is at all busy, you may want to avoid a crosswind pattern entry due to the possibility of cutting off other traffic in the pattern as you make the turn from crosswind to downwind. Additionally, as you make the turn to downwind you will lose some visibility of traffic while you are turning away from them. These factors contribute to the fact that crosswind pattern entries should be used only when traffic volumes permit a safe entry.

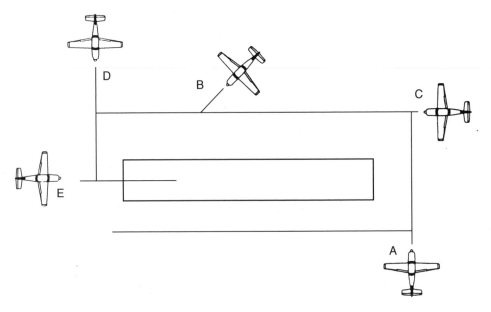

Fig. 6-4 *Pattern entry points.*

Downwind Entry

Downwind entry into the pattern is the most common entry point for uncontrolled fields. There are two types of downwind entry listed in Figure 6-4, points B and C. Of the two entry types, B is the preferred method for reasons of safety and traffic flow and is the one most recommended by FAA publications I have reviewed. Position B is known as a 45-degree downwind entry and provides you with the best view of traffic in the pattern as you approach the downwind leg on a 45-degree angle. This method also provides for a relatively easy turn from the entry angle to the downwind leg and allows you to make any wind correct without an excessively steep bank. Remember, as a rule of thumb you should not exceed banks of 30 degrees in the pattern, and the 45-degree downwind entry allows you to easily remain below this bank angle in most cases.

As you approach the pattern from position B, you should be scanning behind you on the downwind leg to assure that you are not cutting in front of other planes that may already be on downwind. You will also be able to see aircraft on base and downwind legs from this point, giving you the ability to more easily keep track of aircraft in the pattern. This is where listening to the radio during your approach to the airport can give you an idea of where to look for other planes as you fly this pattern approach.

An alternative to the downwind approach using a 45-degree angle is the straight in to downwind approach. Position C shows an airplane flying straight into the downwind leg. While this approach may make an efficient entry point, it can be more difficult to see other aircraft and maintain adequate separation. If you use this point of entry to the pattern, be sure to scan the pattern for other planes. If someone is entering from a crosswind entry point, this could present a possible safety hazard.

Base Entry

Position D in Figure 6-4 illustrates a plane entering the pattern on a base leg. Once again, this entry point can allow a pilot entering from the correct direction an efficient method of entering the flow of traffic. Like other entry points, it can pose problems for the pilot in maintaining proper spacing with other planes in the pattern. In this particular type of pattern entry, you position yourself to fly onto the base leg at the correct distance from the runway. You then make the turn from base to final at the correct position. The challenges with a base-leg entry include maintaining visual contact with other planes in the pattern, correctly estimating the correct distance from the runway, and flying into the leg at the correct altitude. Later in this chapter we will cover knowing when to turn from downwind to base leg, but remember that you will be losing altitude while you are on downwind after you reduce power. Not using the downwind leg to help you set up for the correct position to make the turn from base to final, you may find yourself too high or too low as you turn from base to final. You will also want to be very careful not to cut in front of other planes that are on downwind when you enter the pattern on the base leg; this not only poses a safety hazard but also does not promote a "good citizen" mentality when flying with other airplanes. Unless you know the airport, the traffic in the pattern, how your airplane flies, and what the correct altitudes to use are, flying into the pattern from the base leg may not be the safest pattern approach in many situations. The idea of saving a few minutes by cutting into the pattern on a base leg may seem like a good idea, but not seeing another plane in the pattern could pose problems that will cost you much more than a few minutes.

Final Entry

Final entry to the pattern, also known as a straight-in approach, is depicted as position E in Figure 6-4. To make a straight-in approach, the pilot approaches the airport from a direction that allows him or her to go right to the final leg of the approach pattern. This approach can save time and fuel, much as the base leg approach can, but it poses many of the same problems. Maintaining proper spacing and visual contact with other aircraft can be particularly difficult when you make a long, straight-in approach. It is also tougher for other airplanes to keep you in sight. And like the base leg approach, flying a straight-in approach can make it more difficult to determine the correct altitudes, rates of descent, and airspeeds. The straight-in approach should be used only when conditions and traffic permit.

If you should happen to fly a straight-in approach, you will want to make sure the plane is in landing configuration at least one-half mile from the runway. You should also be at pattern altitude somewhere between one to one-half mile from the runway, and slow the airplane to initial approach speeds and power settings. As the airplane continues the approach, you will need to establish a normal rate of descent, adjusting as necessary to maintain the correct glideslope for the touchdown point. Once you have set yourself up on final, the approach will be flown as if you have started it from any other point in the pattern.

Since you may be executing the approach at altitudes and from a position that you may not use during normal patterns, you will need to know the surrounding area and be aware of any potential hazards such as towers, power lines, or other safety problems. You may also need to be more aware of traffic that may not be in the pattern of the airport you are landing at. Since you are not in the normal approach pattern, other traffic arriving at or departing the airport may not be expecting you at approach altitudes in that airspace. If you do execute a straight-in approach, be sure to exercise caution to assure the safety of the flight.

This concludes the section on pattern entry. In the next section we will review pattern exit options and the methods that you should consider as you leave an airport after you take off.

Pattern Exits

As you probably have already guessed, there are a number of standard pattern exits that you should be aware of and use to improve aircraft separation and visibility. In this section we will discuss each of those options. Like pattern entry points, this is a guideline on possible ways to exit the pattern. Each situation may require that you adapt your path through the pattern to meet a given situation. Let's take a look at the standard departure options.

Crosswind Exit

Position A in Figure 6-5 illustrates the path your plane would fly during a crosswind exit. In this pattern exit you will climb to a safe altitude before initiating a 90-degree turn, in this example to the left. For safety's sake you should attain approximately 500 feet AGL before beginning the turn to avoid hazards such as buildings, towers, or other structures that might pose a risk of impact if you begin the turn at too low an altitude. The turn to crosswind should not exceed 30 degrees bank attitude for reasons we have previously discussed in this chapter.

The crosswind exit allows you to leave the flow of the pattern with minor potential disruption of the traffic pattern. Remember to look for other traffic as you make your turn to crosswind, both to your left and right since you are crossing the downwind leg of the pattern in this situation. While you want to be at 500 feet AGL before you begin the turn, you do not want to be at pattern altitude due to the potential conflict with other traffic in the pattern as you cross the downwind track.

Position B shows another option that is a minor variation to the crosswind exit, a 45-degree exit. In this pattern exit the plane is turned to a heading 45 degrees from the runway heading to move it out of the way of other traffic taking off behind you, and this avoids a perpendicular path across the downwind path of the pattern. After achieving safe altitude, the plane can be turned 45 degrees. The turn can be at a shallow bank, allowing less loss of lift and easier climb during the turn. This pattern departure also makes it easier to see other aircraft that might be on the downwind leg of the pattern as you scan ahead of the airplane during the climb after the turn, but be certain to scan the entire airspace around you as you climb out. Under the right circumstances you will find that using the 45-degree departure can make for more efficient pattern exit and, depending on your heading, a more direct route to your destination.

Right-Turn Departure

Position C in Figure 6-5 shows a right turn exit, in this example exiting a left-handed pattern. The main advantage of a right-turn departure is in saving time to get to your departure course. Under the right situations, making a left-handed pattern exit can take a

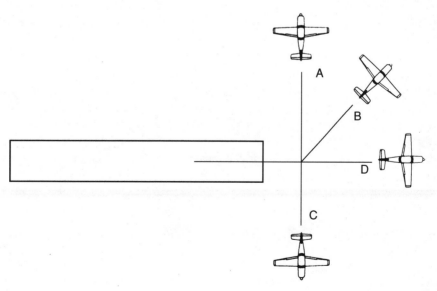

Fig. 6-5 *Pattern exit points.*

relatively large amount of time and distance to get to your on-course heading. In those situations, making a 90-degree turn to the right can help cut down the time and distance you must fly to arrive at your desired heading.

Like the crosswind departure, there are a number of problems associated with using a right turn out from a left-handed pattern. First, if there is traffic entering the pattern on a crosswind or upwind leg, you will be head-on to it, making it more difficult to see the approaching traffic. Since planes entering on those legs are not expecting you to exit in that direction, they may also have a problem spotting you as you head in their direction. As you plan for a right-handed pattern departure before takeoff, you should make yourself aware of other traffic in the pattern. If the flow of traffic is heavy, or there is a potential for conflict with other traffic, you should opt for another pattern exit. Listen to the radio as you start the airplane and taxi out to learn what planes are doing in the pattern as they approach the airport. Remember, keep safety as your highest priority and don't let the desire to save a few minutes put you in a potentially dangerous situation.

Straight-Out Exit

The last pattern exit we will discuss is the straight-out exit. Essentially, in this pattern exit you will continue to fly the runway heading as you climb out after takeoff. Like the other pattern exits, this one is most often used when this heading best coincides with your on-course heading. It can also be an option when obstructions on either side of the flight path might pose hazards if you begin a turn from the runway heading. Airports located in valleys, or cut out of a forest, are an example of when you might need to continue to fly the runway heading until you have achieved sufficient altitude. One consideration of whether you should execute a straight-out departure is what traffic that might be taking off behind you is like. If you are flying a slow-moving airplane and there are higher-performance airplanes behind you, there may be a chance of conflict if the other planes overtake you. Being aware of other aircraft can help you determine if executing a straight-out departure might pose a safety hazard. In this situation it may be wise to make a slight turn to the left or right to help clear the departure path along the runway heading.

Each of the pattern exits we have discussed in this section may need to be adapted to fit a given airport, pattern, or traffic situation. Knowing what options you have can make it safer and more efficient for you and your passengers as you exit the pattern. You should also keep in mind that at tower-controlled fields, tower staff will direct you regarding the headings they want you to fly as you leave the airport. They will also direct you on altitudes you should use and when to make turns to headings that allow for smooth flow of traffic around the vicinity of the airport.

LANDING TECHNIQUES

Each of the topics we have covered in this book has built in knowledge we previously covered. In this section we are going to discuss various techniques you will use during landings that will in part be based on topics we have discussed previously in the book, while introducing new concepts that will become the foundation for improving your landings. Among the new areas we will discuss are the key position in the pattern, picking your landing spot, the flare, or roundout, and touchdown on the runway. In following sections we will review various types of landings, and the information we cover in this section will be crucial in being able to consistently and safely practice each of those landing scenarios. Let's begin with airspeed control and power settings during landings.

Airspeed Control and Power Settings

We have already discussed airspeed and rate of climb or descent control in previous chapters. Now we will apply those control techniques to landings as well. Student pilots who are learning landings frequently make the mistake of equating power setting with airspeed control, and the elevator with their rate of descent. We already know that this is, in fact, the opposite of what actually takes place. To successfully fly a constant glideslope and land on the desired touchdown spot on the runway, you need to maintain a consistent airspeed and rate of descent. This means that you will need to stabilize your airspeed early in the approach at the correct values for the plane you are flying, and hold a constant power setting to aid in flying a rate of descent, and therefore glideslope, that does not vary during the approach.

At the appropriate point on downwind, you will reduce power to the approach and descent setting recommended by the manufacturer. For most single-engine general-aviation planes, this is frequently around 1500 to 1700 r.p.m., but may vary for your aircraft. If you vary the power setting up and down during the approach, this will have the effect of increasing and decreasing the rate of descent as you fly, resulting in a glideslope that is not constant. In order to accurately touch down on your target landing point, you will need to fly a glideslope that has a constant angle. This means you need to maintain a glideslope that does not vary in its angle throughout the landing approach, and to accomplish this you must hold a constant power setting until the point you bring it to idle.

If you find that you are above the glideslope, you should gradually reduce power, allowing a greater rate of descent. As you bring the power back, you will need to adjust the plane's pitch to maintain the correct airspeed. The greater rate of descent will allow you to intercept the necessary glideslope. You will need to anticipate the plane intercepting the correct glideslope, increase power at the appropriate time to the correct power setting to arrest the rate of descent and achieve a glideslope that is parallel to the desired glideslope. At the other end of the spectrum, if you are below the correct glideslope you must increase power to reduce the rate of descent in order to intersect the correct path to the runway. As you increase power, you will need to increase backpressure on the elevator to maintain your airspeed. The combination of increasing power and elevator backpressure has the effect of reducing the rate of descent. As you achieve the correct glideslope, you will need to anticipate reducing power to the correct value to maintain the proper rate of descent. Remember, with the change in power setting you will again need to change the pitch of the plane to hold your approach airspeed.

As you become proficient at setting up for landings, you will find that you will minimize changes in power settings and airspeeds, allowing you to achieve very consistent glideslopes. This consistency is key to becoming competent at picking out your touchdown point on the runway, setting up for the approach, and landing within feet of that spot on the runway. To me this is one of the best ways to measure how well a pilot understands how to fly an airplane. All of the concepts of the basics of flying are inherent in making good, consistent landings, and this is why one of my flight instructors from many years ago said he could determine how good pilots are by just flying around the pattern with them. Learn how to maintain a constant glideslope and airspeed as you execute your landings, and how to make adjustments during your approach that do not result in your moving above and below the glideslope as you increase and decrease power and vainly chase the correct rate of descent.

Picking Your Touchdown Point

Previously in this chapter we have discussed flying the pattern and each of the legs in it. Now we are going to begin to learn when to reduce our power to approach settings, and when to make the turn from the downwind leg to base. Knowing when to perform each of these tasks is one of the first steps in learning to fly a smooth, consistent approach to landing. Many students learn to rely on landmarks around the airport they learn to fly at to cue them as to when to reduce power or make a turn. But this can be a problem for these new pilots as they begin to fly to other airports where those landmarks are not available. By learning to perform power setting changes and turns based on your position from the runway, you can avoid the problems of not having the grove of trees or the barn you normally use as a reference unavailable at airports you visit.

In addition, we will discuss when to lower your flaps as you fly through the approach. After you finish this section, you will have a good understanding of when to take each step as you fly through the pattern, helping you become more consistent in picking out your landing zone, making your turns at appropriate distances from the runway, and touching down near the landing spot you picked out while on downwind.

THE 45-DEGREE POINT

The textbook approach to landing has a number of steps that must be completed at appropriate points from the time you enter downwind until the plane touches down on the runway. In our example for this section, we will assume that you have entered downwind at the published altitude, have completed the prelanding checklist and gone through the GUMP (gas, undercarriage, mixture, prop) checklist, and are flying at pattern airspeeds. As you come abeam your target touchdown point while on the downwind leg, you will reduce power to the setting recommended by the manufacturer. In this example we will use 1700 r.p.m. as the power setting of choice.

Once power has been reduced, you will need to slow the plane to the approach airspeed while initially maintaining altitude. This is accomplished by easing back on the elevator control until the target airspeed is achieved. For this example we will use 75 knots as our approach speed. After we have slowed the plane, we then begin our descent, maintaining the 75-knot target. Remember, if we are slow we must lower the nose slightly to increase airspeed, and if our airspeed is high we need to pull the nose up just enough to reduce the airspeed to the correct value. With a little practice you will be able to hold the airspeed constant with minor elevator inputs throughout the approach. Our power is now set at 1700 r.p.m., we have 75 knots on the airspeed indicator, and our rate of descent is about 500 feet per minute (fpm). At this point in the approach we put on our first notch of flaps to help maintain our airspeed and set the rate of descent. Be prepared for a change in pitch as the flaps are applied; many aircraft will have a tendency to pitch nose down with each increment of flaps that are applied. Hold the airspeed constant at 76 knots through elevator control inputs and then trim the elevator to reduce control pressures to neutral. Trimming the plane is an important step that many pilots forget, but it can help you reduce your workload, physical and mental fatigue, and help stabilize the airplane's pitch and airspeed if you should become momentarily distracted. These first few steps should take place within ten to fifteen seconds after the power is reduced.

Now that we have the plane flying along with the airspeed and power at correct settings, the first notch of flaps applied, and our prelanding checklists completed, we need to think about when we should make our turn to base. Figure 6-6 shows a plane flying on downwind. When the plane reaches a point where it forms a 45-degree angle with the touchdown point, you will want to execute your turn to the base leg of the approach. The position where the 45-degree angle takes place is often referred to as the *key point*, and it is necessary to look back over your shoulder, the left in this example, to verify when you have reached the key point. In the case of a right-handed pattern, you will still use a key position, but you will be looking back to the right to see when the correct position has been reached.

Notice that this point will hold true at any airport, and the use of ground references no longer is necessary to figure out when to make the turn from downwind to base. Make sure your bank angles do not exceed 30 degrees during the turn. If workloads permit, you should announce over the radio that you are turning to base. The reason for letting other pilots know you are turning to base while in the turn is that it is often easier for planes behind you in the pattern to see your plane while it is banked. The wings present a larger and easier-to-see target when they are banked towards the traffic behind you. While on base your airspeed and power setting should remain constant, and after you roll out on to

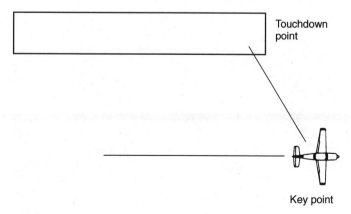

Fig. 6-6 *Key point.*

the base leg you can begin to figure out if you are on the correct glideslope for your planned touchdown point.

As the plane continues to descend on base, notice the position of the landing zone on the windscreen. If the touchdown point rises in the window, your glideslope is too steep and you will land short of your mark. If the touchdown point sinks on the window, your glideslope is too shallow and you will land past your target. It takes a little practice to begin to quickly figure out which way the touchdown point is moving in relation to the windscreen, but the ability to figure this out is key to being able to adjust your glideslope to land at the correct point on the runway. If you find the touchdown point is moving up on the window, you will need to increase power to reduce the rate of descent and increase elevator backpressure to maintain 75 knots of airspeed. Once you have put the touchdown point at the correct position, you will then reduce power slightly to hold the point steady, again adjusting elevator pressure to maintain the correct airspeed. If the touchdown zone is moving down the window, you will want to reduce power slightly, maintaining the correct airspeed, to increase your rate of descent. Again, once you have put the landing spot back at the correct position on the window, you will need to increase power to hold the proper glideslope. You can see how important it is to use power and airspeed in harmony to hold a steady glideslope.

The touchdown point will hold a constant position on the window if you have the correct glideslope set up during the approach. Experience will help you determine what the proper position for the touchdown point is on the window, and the position will vary from plane to plane. But if you get into the habit of beginning to observe the relative movement of your intended area of landing on the runway, and adjusting while you are still on the base leg, you will find it easier to land just where you were aiming.

Now that we are on base, we will want to extend a second notch of flaps and begin to reduce our airspeed. For our example we will slow the airplane to 65 knots, resetting trim after we have extended flaps and slowed the airspeed to 65 knots. As with the first notch of flaps, be prepared for any pitch changes as the flaps are extended. With the reduced airspeed you will notice a reduction in control effectiveness. You will need to adapt your use of the controls as the airspeed slows throughout the approach. As soon as we turn onto base, we should begin planning our turn to final. Distance from the runway, the speed of the plane, and the direction and speed of the wind are critical factors in when we should begin the turn. Like everything in flying, the more experience we get, the better we will become at judging when to begin the turn and how steep to make the bank during the turn. Once you begin to make the turn to final approach, you may need to adjust the bank to compensate for the wind's effect on your ground track. When the wind is to your back during the base leg, it is not uncommon for students to begin with too shallow a bank, underestimating how much speed the wind adds to the ground speed. Once the realization that they are not going to end up aligned with the runway sinks in, pilots then either bank too steeply in an attempt to salvage the turn, or end up well to one side of the runway centerline and must work to get the plane back where it belongs.

As you roll out on final, you should again monitor the position of the intended touchdown point on the windscreen, watching for any movement up or down. The corrections while on final are the same as those we discussed on the base leg. Many pilots find it easier to visualize what the touchdown point is doing once they are actually line up on final approach as compared to the base leg. In our example we will now add the third notch of flaps and slow the plane to 60 knots airspeed, adjusting trim to reduce elevator control pressures after we extend the notch of flaps. Monitor the movement of the touchdown point after the third notch of flaps are added. The flaps will increase the drag and increase the rate of descent while preventing the plane from gaining too much airspeed. Make adjustments in power and pitch as necessary to maintain the correct glideslope. Figure 6-7A, B, and C show three views of possible approaches while on final. 6-7A shows the runway sliding down the windshield, the plane higher than it should be on the glideslope. Figure 6-7B shows the plane on the correct glideslope, with the landing zone stable in its position on the windscreen. Finally, Figure 6-7C shows the view you might see when the plane is below the glideslope, the touchdown point moving up the windshield.

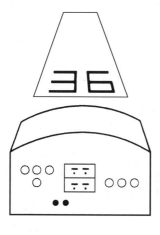

Fig. 6-7A *High final approach.*

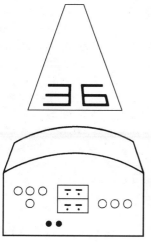

Fig. 6-7B *Normal final approach.*

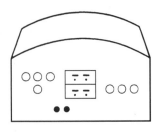

Fig. 6-7C *Low final approach.*

Once you are assured of making the runway, for many single-engine aircraft you can close the throttle to an idle. For some aircraft it may be necessary to maintain power until you are actually about to flare, but the practice of carrying power is normally used on heavier aircraft that may need the extra power to help maintain elevator effectiveness or to avoid nose-down pitching moments that can take place if power is reduced to idle. During approach, use the power setting that is recommended by the aircraft manufacturer, however. I like to teach pilots to make a power-off approach for one major reason. If they can make it to the runway with the power off, then they don't have to worry about losing an engine at that point and landing short of the runway.

In fact, I try to fly all approaches with the power reduced to idle when I am abeam the touchdown point on downwind. If I plan on the approach not having any power, and fly it to make sure I will make the runway no matter what, I do not need to be concerned with losing the engine at any time during the approach. To many of you, this may seem a bit strong on worrying about losing the engine, because the engines we fly with are so reliable. But a few years ago while giving a sightseeing tour around a city near where I live, I was descending toward the tower-controlled field. It was a slow day for the tower and they had cleared me to land. As I reached the altitude I wanted to level off at until I was closer to the airport, I advanced the throttle and found much to my surprise that the engine was producing no power. The propeller was windmilling, so my passengers were not aware that anything was wrong. I set up for best glide speeds, then did all the steps you are supposed to do (carb heat, change fuel, etc.), but nothing had any affect on the engine. The wind was pretty strong that day, and I wasn't sure I could make it to the airport boundaries, let alone a runway, but keeping a highway as my alternate landing spot, I made as straight a line as possible to the airport.

After what seemed like hours I reached a point where I knew I could land at the airport, but not on a runway. In a very anticlimactic event, I executed a soft-field landing in the grass and rolled to a stop. The problem turned out to be a sliver of brass that had wedged between the needle and seat in the carburetor and had flooded the engine. The mechanics found the problems when they took the carburetor apart the next day. They reassembled the airplane, and I test flew it without any problems. Ever since the early days of my flight training, my instructor taught me to make approaches power off whenever possible. I now understand very clearly why this is important. While this is not always possible due to traffic flow or tower instructions, get in the habit of considering what you would do if you ever did lose power during the approach. But for our example we are now in position on final approach, with the plane in landing configuration. From this point we need to get ready for the landing itself, the topic of our next section.

FLARE AND TOUCHDOWN

As you continue down final approach, you will need to slow the airplane to the "over the fence" speed as you near the surface of the runway. The transition from final approach to touchdown involves several steps that will help you reduce the rate of descent, level the airplane into the flared attitude, and then touch down. One of the most critical factors in making the transition from final to the flared position is vision. You need to be able to accurately judge your height above the runway to execute the flare, and vision is one of the most important components of being able to determine your height.

Your ability to judge your altitude will improve with experience, but there are several criteria you can use to aid in accurately determining your height above the runway. We previously mentioned vision, and part of this is where you focus during the landing. Many pilots have a tendency to focus too near the airplane, or to focus in just one spot. Figure 6-8 shows the results of focusing your view too close to the plane. Your vision can become blurred as you approach the ground, making it more difficult to judge your altitude. Focusing in just one spot can also prevent you from getting a good feel for your height. Instead, you should move your field of vision from the far end of the runway to in closer to the plane throughout the approach. The higher the airspeed your plane lands at, the further out in front of the plane it becomes necessary to look. As the plane slows, you can gradually bring the points you are focusing on in closer to the plane.

If you focus too close to the front of the plane, you will have a tendency to think you are lower than you really are and to drop the airplane in during the flare, as opposed to a nice, smooth touchdown. If you keep your focus too far in front of the plane, you will have a tendency to think you are higher than you actually are, and you may not flare soon enough. In tricycle-gear aircraft, this can result in your landing on the nosewheel first, or wheelbarrowing down the runway.

I also like to use peripheral vision to help judge my height as I begin the flare. By using the position of the edges of the runway on the side posts of the windscreen, I can add to my ability to accurately determine my altitude above the runway. The next time you turn on to the runway for takeoff, get a feel for about where the edges of the runway are located on the side posts. As you land, use your peripheral vision to keep track of the

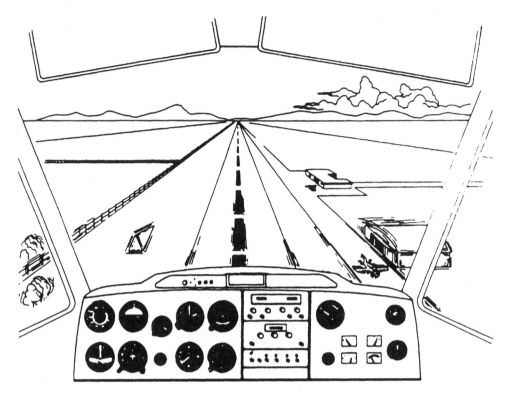

Fig. 6-8 *Focusing too close to plane.*

runway edge as you begin to flare. You will notice that this can help you better judge your altitude as you get down to the last few inches above the runway, allowing you to smoothly descend to touchdown.

The most important factor in learning to consistently judge your altitude is practice. Every plane has a different feel, and look, as you land. Get used to what the picture is as you land the plane you fly, then compare it to other aircraft you have been in. The more you practice and the better you become at critiquing your landings, the better and smoother you will become at landing. Now that we have discussed vision during transition from final to touchdown, let's cover the actual landing.

Since most pilots fly tricycle-gear aircraft, we will focus our attention on that type of airplane. However, if you fly tailwheel planes, you will find that many of the concepts apply to those planes as well. The key difference is that tricycle-gear airplanes should be landed on the main landing gear with the nosewheel off the ground, while taildraggers can be landed in either the three-point position or be "wheeled" on, landing on the main landing gear with the tailwheel held off the runway. In either case, you need to transition the plane from the nose-down attitude that was established during the approach to a flared attitude, allowing the plane to gently settle to the runway. That transition to flare will be the topic of this discussion.

While on final we are still in a nose-low attitude, with power at the correct setting and flaps and gear extended. The airplane has been slowed to the manufacturer's recommended airspeed for final approach, and we are continuing toward our landing zone. The touchdown area is stabilized on the windscreen, indicating that our glideslope is correct. As we cross over the end of the runway, or as some refer to "over the fence," we must begin slowing our rate of descent and level off just above the runway. To accomplish this, we must ease back on the elevator control, raising the nose of the plane. This has two effects. It slows the rate of descent and begins to bleed off airspeed, both necessary components for a full stall landing in a tricycle-gear airplane. The altitude where you will begin to raise the nose will vary for each airplane and is dependent on your airspeed, the rate of descent, and the responsiveness of the airplane's flight controls. The Pitts Special I fly has a very high rate of descent and approach speed, and I begin to ease the control stick back when I'm about fifty feet from the ground. But a Cessna 152 can descend closer, depending on conditions, before you need to begin the flare.

The flare, or the action of raising the nose of the plane during the transition from final to landing that we have just discussed, is also known as the *roundout*. Flaring the plane at the proper altitude, and with the proper application of elevator backpressure, becomes easier with experience. Student pilots often begin by flaring either much too high due to concerns about landing too hard, or flaring too low, not recognizing the actual height above the runway. The idea is to not pull back at too high an altitude or too quickly, and not to let the plane hit the runway nose-wheel first because the elevator control was not eased back soon enough. The timing of the flare can be closely tied to having the correct vision of the landing and being able to accurately judge your altitude. Now we can see why being able to determine your altitude is so important to a smooth landing.

Pulling back on the elevator at too high an altitude can cause you to slow and stall while you are still too far above the surface of the runway, while pulling back too rapidly on the elevator control can cause the plane to gain altitude and lose airspeed at the same time. Like beginning the flare too high, this can cause the plane to stall while well above the runway and drop onto the runway. In extreme cases this can result in severe damage to the aircraft and injury to you and your passengers.

Figure 6-9 shows the correct application of elevator at the correct altitude, with the plane transitioning from the nose-down attitude to the flared, nose-high attitude needed for touchdown. In the ideal landing scenario, you will be just at stall speed as the main gear touch down on the runway. However, it is not uncommon for pilots who are not proficient in full-stall landings to touch down with speeds well above the stall speed. The major consideration is holding the right nose-up attitude once you are in the flared position, and letting the airspeed bleed off as the plane settles to the runway. How much elevator, how quickly you will need to apply it, and the correct nose-up pitch attitude needed will depend on the plane you are flying. Some aircraft, especially those with retractable gear, tend to be more nose heavy and may have a heavier feel on the elevators. Trim can help reduce some of this heaviness, but you should anticipate the need to apply pressure early enough to avoid any problems with not being in the correct attitude for landing. Once the plane does touch down on the main gear, do not drop the nosewheel onto the

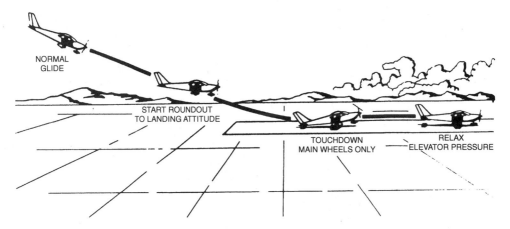

NORMAL
GLIDE

START ROUNDOUT
TO LANDING ATTITUDE

TOUCHDOWN
MAIN WHEELS ONLY

RELAX
ELEVATOR PRESSURE

Fig. 6-9 *Roundout to touchdown.*

runway. Instead, slowly lower the nose to avoid any unnecessary impact with the runway. If runway length permits, you can actually hold the nosewheel off the ground for an extended distance. Holding the plane in this attitude can help slow the plane through aerodynamic breaking and save on break wear.

Let's review the transition from final approach, through the flare and touchdown, with an example to make sure that each of the steps is clear. As we descend on final, the airplane should be at the recommended airspeed for final approach with gear and flaps extended. As we approach within twenty to thirty feet of the surface, we begin to ease back on the elevator control, raising the nose of the plane and reducing airspeed. As the plane slows, we should shift the area of the runway we are focusing on close to the airplane, but don't fixate on just one spot. Look around ahead of the aircraft and get a feel for your altitude. If power has not been reduced to idle, this may be a good time to pull it all the way back, depending on the recommendations of the manufacturer. As the plane loses airspeed, you will need to add additional elevator backpressure to maintain the correct nose-high attitude for touchdown. If you have judged the height of the flare correctly, the plane will settle the last few inches to the runway's surface as the remaining speed bleeds off, touching down on the main landing gear. Due to the nose-high attitude, your view of the runway ahead may be partially blocked, but most tricycle-gear aircraft have good forward visibility even in the flared position. Hold the nose off the runway, slowly easing forward on the elevator control to let the nose gear settle to the ground in a controlled manner. Initial directional control can be maintained via the rudder, but as the plane slows it may take a combination of nosewheel steering and differential braking to control the steering of the plane.

At this point keep flying the airplane! Just because the plane is on the ground does not mean you can relax and think that nothing can go wrong. Too many accidents are caused each year because the pilot's attention switched from maintaining control of the plane to talking to passengers or changing radio frequencies or other tasks. You still need to compensate for wind, watch for potential trouble spots on the runway, and make sure you are headed in the right direction. Remember to fly the plane from the time the propeller starts turning until it stops at the end of a flight.

GROUND EFFECT

An area that seems to be shrouded in misconception is ground effect, and many pilots do not understand what it is, or how it affects their plane during both takeoff and landing. A common perception is that ground effect is a cushion of air that builds up between the wings of the plane and the runway, causing the plane to float down the runway. In this section we will discuss what ground effect actually is and how it affects you when you fly.

Earlier in the book, we discussed how lift is created, but let's get a bit more in depth for our ground effect discussion. Figure 6-10 shows the changes in airflow that take place as a plane flies in close proximity to the runway. As the airflow moves past the trailing edge of the wing, a downward angle is imparted to it. When you are generating lift within close proximity to the ground, the flow of air at the trailing edge of the wing is affected and does not bend downward as sharply. The effect on the airflow becomes greatest when the plane is less than one-quarter of the wing span's distance above the surface. In fact, induced drag, or drag generated as a result of lift generation, is reduced by 47.6% when within one-tenth the distance of the wing span to the ground. This value is reduced to 23.5% at one-quarter span and drops to only about 1.4% at one full span's distance. The vertical component of the airflow around the wing is affected by the runway, reducing the wing's downwash, upwash, and wingtip vortices. As a result there is a smaller rearward lift component and less induced drag generated by the wing.

A number of beneficial efficiencies are achieved as a result of this change in airflow and reduction in induced drag. First, to generate the same lift coefficient, a wing in ground effect will require a lower angle of attack than the same wing not in ground effect. There is also a reduction in the amount of thrust required to maintain a given aircraft velocity. This higher coefficient of lift, combined with lower thrust requirements, are what actually cause a plane to float down the runway when the pilot lands carrying too much speed or extra power, without the "cushion of air" we mentioned earlier.

Ground effect also generates an increase in the local air pressure at the static source, causing a lower-than-normal indicated airspeed and altitude. During takeoff this can make it seem as though the plane is able to fly at lower-than-normal airspeeds. As the

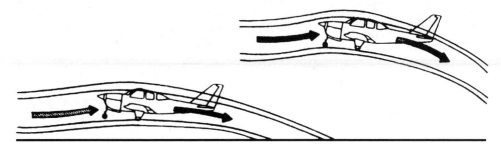

Fig. 6-10 *Ground effect.*

plane leaves ground effect after takeoff, several things take place. Due to the increase in induced drag and the reduction of the coefficient of lift, the wing will require an increase in the angle of attack to generate the equivalent amount of lift. Additional thrust will also be needed as a result of the increase in induced drag. Finally, the change in air pressure around the static source will cause the airspeed indicator to register an increase in the indicated airspeed. There will also be a decrease in the aircraft's stability, along with a nose-up change in moment (*Flight Training Handbook*, p. 271).

As you can see, there is no magic associated with ground effect. There are very measurable changes in airflow which result in a reduction of induced drag and improved lift coefficients. If you understand how ground effect changes the manner in which your plane flies, you can use this knowledge to become a safer pilot during takeoffs and landings. We have all heard stories about one pilot or another who managed to get their plane, usually loaded above the allowed maximum gross weight, off the ground while in ground effect, but they were unable to climb after liftoff. This is why this can occasionally happen and why you should never trust luck as you load your airplane. Fly it within limits recommended by the manufacturer, and you will reduce your chances for trouble.

NIGHT LANDINGS

Night landings are one of my favorite landing situations. This is where you get to really test your ability to visualize the runway and your altitude above it and work on the timing of the flare so that you still touch down right where you want to with a nice, gentle squeak of the tires. At night you lose many of the visual references that are available during the day—ground reference points, trees, fences and buildings that you might use to help judge your altitude above the ground. Instead, you must fly your approach using the movement of the runway lights to tell you whether you are above or below the glideslope, on the centerline of the runway, and where you are touching down on the runway. Because you must execute the approach based solely on the runway lights and your relative position from them, your skills at determining your altitude, the timing of the flare, and directional control of the airplane are put to the test. However, with a little practice, night landings are just as easy to perform as safely and competently as landing during the daytime. In this section we will cover a number of techniques you may find helpful in becoming more proficient, and confident, in your night landings.

Approach Planning

You will notice immediately that when you fly at night the visual queues you may have come to rely on for making a turn, reducing your engine power, or any other task is no longer readily available. For that reason you must be proficient at flying the approach in the manner outlined in the previous section through the use of references that are related to the runway, or the runway lights at night. You will set up for downwind just as you would during the day, holding the same altitudes and distances from the runway while on the downwind leg. At night it will be necessary to pick a runway light as the marker for your touchdown point, instead of the centerline stripe you might use during the day. Distance from the runway can be determined in the usual way, using the runway lights' position relative to your wing or strut. At night the lights will line up in about the same place as the edge of the runway would during the day.

There are a number of different factors that can affect when you should turn on your landing light. At an airport that has much traffic in the pattern you should turn your landing light on as you get near the airport's vicinity to improve the chances of other traffic being able to see you. In lower traffic areas you may want to wait until you are on downwind before turning the landing light on to help conserve its life. As you come abeam the runway lights you are using for the touchdown point bring your power back just as you would during a day approach, setting up the correct airspeed and descent values.

Once you have pulled the power back, trimmed the plane, and set the appropriate amount of flaps, you will continue on the downwind heading until you reach the key position we discussed previously. At that point begin your turn to base, letting other traffic know what you are doing over the radio. Continue flying the approach, watching the touchdown point for movement up or down on the windscreen. As with a day approach, if it is moving down the windshield you will overshoot the landing area and you will need to reduce power to increase your rate of descent until you intercept the correct glideslope. If the runway lights you want to land next to are moving up the windshield, you are going to land short of them. Increase power to intercept the correct glideslope. You can begin to watch the movement of your touchdown point as soon as you turn base to give you as much time as possible to correct for any glideslope problems.

As you fly the base leg, add flaps and maintain airspeeds just as you would during the day. You may find that night approaches make it more difficult to maintain your attitude visually from the horizon. Cross-check your external attitude reference with the instruments inside the plane. The artificial horizon, vertical speed indicator, and airspeed indicator become very important instruments during night approaches. Make sure you don't bury your head in the cockpit, however, and ignore looking out the windows of the plane. If the runway you are landing on has VASI (Visual Approach Slope Indicator), you can also use this as a reference for touching down in the correct spot on the runway. At this point we are on final approach, our airspeed is set, we've gone through the prelanding checklists, and it's time to think about when to begin the flare.

Height Judgment/Flare

During night landings you will have a limited field of view as you get close to the runway. Day landings allow you to see an almost unrestricted view of the runway and airport, but the runway illuminated by the landing light will be the only area you will have a clear view of. On very bright, moonlit nights you may gain some additional illumination, but this is often not the case. This mean you must learn how to judge your height above the runway based on other references.

The first references you can use are the runway lights themselves. In much the same way we discussed using the edges of the runway and their position along the side posts of the windows, the runway lights can be used in a similar manner. Using peripheral vision, you can note the runway lights moving up on the side posts as you descend closer to the runway. As you descend low enough that the runway becomes illuminated by your landing lights, you have reached a point that you can use the runway itself as a visual reference.

Some texts recommend waiting until you see the tire marks on the runway before beginning the flare. Figure 6-11 shows a view that is typical of night approaches, the runway outlined by the runway lights until you descend low enough to illuminate the runway directly in front of the plane with the landing light. Depending on the type of plane you fly, this procedure for flaring might work. Another recommendation is to use the lights at the far end of the runway as a reference point. When those lights appear to rise higher than the airplane, begin the flare. This may also work under some circumstances, but if the runway is long this technique may pose problems.

Practicing night landings is the best way to become proficient at them and to develop a technique that works for you and the plane you fly. If you are at all rusty at night landing, or uncomfortable with them, find a qualified flight instructor to help you become safer and more proficient. With enough practice you will find that it is possible to judge your height above the runway without any landing light at all. While I do not recommend this for every pilot, knowing that you can still land at night, without a landing light, is a great confidence booster. Also keep in mind that when making night landings you will need to be prepared to compensate for the effect of wind. We will be discussing crosswind-landing techniques later in this chapter, and they will apply equally to day or night landings. The biggest challenge with night landings is that you have fewer references to figure out what the wind is actually doing to your approach. But nothing feels better than shooting a night approach and feeling that gentle touch of the tires on the runway. Now let's take a look at some of the common takeoffs and landings you will need to master.

NORMAL TAKEOFFS AND LANDINGS

So far in this chapter we have reviewed the mechanics surrounding takeoffs and landings, how to enter and exit the pattern, the techniques to use to hit your landing zone, and a number of other topics that are all part of learning to be proficient, safe, and consistent at your takeoffs and landings. Now we are going to look at the specifics for several different takeoff and landing situations, ranging from normal takeoffs and landings to some of the more specialized situations. We will discuss control use, applicable situations to use these takeoff and landing techniques, and common errors pilots tend to make when executing them. Let's begin with normal takeoff and landings.

Normal Takeoff

A normal takeoff can include many takeoff situations and includes the basic takeoff techniques that are part of the other takeoffs we will cover through the rest of this chapter. To begin discussing normal takeoffs, let's walk through an example.

Before any takeoff you will need to complete the preflight inspection and takeoff checklist. The pretakeoff checklist can include many items, but generally includes checking the controls for proper movement, the instruments for proper function, that your fuel

Fig. 6-11 *Night landing.*

is set to the proper tank, that your fuel tanks are filled properly and your mixture is correctly set, that the engine is runup and the mags are checked, that fuel and oil pressures are in the normal ranges, that carburetor heat is working, that your trim and flaps are set for takeoff, and finally that seatbelts and doors are secured. Once you have completed the checklists you will need to obtain takeoff clearance at a controlled field, or notify traffic via the proper frequency at uncontrolled fields. At both controlled and uncontrolled fields, you will want to visually clear the pattern and runway to make sure there is no other traffic that poses a safety hazard BEFORE you taxi onto the runway. Once you are assured that there is no traffic in your vicinity, taxi onto the runway, aligning the airplane with the runway centerline. Do a final check of trim, flap settings, and engine instruments as you finish the turn onto the runway. You will also want to scan down the runway and intersecting runways for any aircraft, airport vehicles, birds, or other potential hazards during your takeoff.

At this point you have assured that the plane is ready to fly, and that the airspace and runway are clear of any traffic. The ailerons, elevator, and rudder should be in the neutral position as the throttle is advanced smoothly to full power. As engine power is applied you will probably need to apply some right rudder to compensate for any left turning tendencies as a result of torque or the other left turning tendencies we have previously discussed. Once the airplane reaches the airspeed recommended for rotation by the manufacturer, ease back on the control yoke, raising the nose gear from the ground. You should be attempting to reach the correct takeoff pitch angle with the rotation. Depending on the airplane you are flying, the plane may continue to accelerate with the main gear still on the runway, lifting off once flying speeds have been reached. With other aircraft you may find that the plane lifts off almost immediately after the nose has been raised.

As the nose is raised, you will probably need to adjust the amount of right rudder being applied. With the increased angle of attack, P-factor becomes stronger and may require additional right rudder to keep the plane tracking straight. After the airplane leaves the runway, establish a pitch angle that gives you the climb speed recommended by the manufacturer, normally V_x or V_y, depending on the situation. During the takeoff roll and liftoff you want to keep the control inputs as small and smooth as possible. Some pilots I have flown with have a tendency to be almost mechanical in their control inputs, not actually feeling what the plane is telling them and adjusting their control inputs to a particular situation. There have been cases of pilot-induced oscillation because the pilot overcontrols initial inputs, then overcompensates, trying to correct for the first problems. The most severe cases of this type of overcontrolling can result in loss of control of the airplane. If you should find yourself in this situation, neutralize the controls and let the plane stabilize, then use less aggressive control inputs to establish the correct flight attitudes. An important factor in situations where you have overcontrolled the airplane is to recognize that the situation exists, then stop making unnecessary inputs until the problem is brought under control.

During the takeoff roll and climb out, ailerons should be in the neutral position except to maintain wings level after takeoff, or as you begin banking the airplane in order to turn onto the correct departure heading. Remember to keep the use of ailerons and rudder coordinated as you compensate for any left-turning tendencies that are present during takeoff. Monitor flight and engine instruments throughout the takeoff, but don't bury your head in the cockpit and ignore what is outside the plane.

A number of errors are common as pilots learn to execute normal takeoffs. One of the first is raising the nose at too slow an airspeed, or adding too much nose-up elevator. In either case the nose of the plane can pop up from the runway before the plane is ready to fly, and the added drag from the nose-high attitude can increase drag and the ground roll necessary for takeoff. Use smooth, controlled inputs to help avoid this situation. Raising the nose to an excessive pitch angle is also a common error until pilots get used to the sight picture they will have during a normal takeoff. Maintaining the nose at too high an attitude could result in the plane stalling, a problem we want to avoid close to the ground. Another error that is actually quite common is not using right rudder during the takeoff roll. When pilots do not compensate for left-turning tendencies, they will invariably drift toward the left side of the runway during the takeoff run, and in severe cases they have actually left the runway. I know of two cases where student pilots ended up flipping the airplane on its back as the plane left the runway due to lack of compensation for left-turning tendencies with right rudder. In each case the pilot was not seriously injured, but the plane was severely damaged. The last error we will discuss is leaving the nosewheel on the runway too long and letting the airspeed build too much before raising the nose. This can cause excessive wear on the nosewheel tire and could result in nosewheel shimmy, which will cause vibrations through the airplane. In severe cases damage to the nose gear or loss of control of the plane could be the result.

Normal Landings

In this section we will discuss landing procedures for tricycle-gear aircraft. We will assume that you have correctly entered the pattern, have completed the prelanding checklist, and that the gear is in the down and locked position. Let's begin with flap settings.

Flap Settings

We already know that flaps are used to help reduce stall speed and allow a greater rate of descent without an excessive increase in airspeed. Both of these factors work to our advantage during the approach. It is necessary to slow the plane down to shorten the landing distance needed and to allow control of the airplane at these reduced airspeeds. The amount of flaps you use during a landing will depend on the conditions during the landing. Available runway, wind speeds and angles, the load you are carrying, and other factors can dictate the amount of flaps you extend.

In many cases only one notch of flaps may be needed, while in others full flaps can be used to your advantage. Particularly when you want to descend over an obstacle near the approach end of the runway, full flaps can help increase the rate of descent to get you down to the runway more quickly and reduce the amount of runway you will need during landing. If high crosswinds are present, it may be necessary to use less flaps to avoid getting so slow that you lose enough control authority to overcome the effects of the wind. When you should apply flaps has been discussed previously in this chapter and can be modified to meet a situation. If you need only one notch of flaps, you may want to wait to apply it until you are established on final approach, while in other cases you may want to apply flaps on the downwind or base legs. In any case, use the setting recommended by your aircraft's manufacturer.

A common error associated with the use of flaps is not anticipating the change in pitch that can accompany the extension of flaps, often resulting in an excessive loss of airspeed. Keep in mind that you will need to adjust your pitch and retrim the airplane after each notch of flaps is extended.

Trim Settings

Setting the trim can help you become more exact in your airspeed control throughout the approach. When the elevator trim is set correctly, you will find that the plane has a tendency to maintain an almost constant airspeed, in calm air, with little pressure on the control yoke. Pilots often neglect to retrim the airplane after power setting or flap changes, increasing the control pressure they must use to hold the correct airspeed. Use the trim to your advantage, allowing it to do the work of reducing the control pressures needed during your approach. With a more constant airspeed during the approach comes a more constant glideslope and a more accurate ability to land at your intended point on the runway.

Some pilots have expressed concern about what trim will do during a go-around, where it is necessary to apply full power and climb. They feel the nose of the plane will rise too quickly, or that by overriding the trim with the elevator they will damage the airplane. If you must execute a go-around, control the pitch angle with the elevator. While this may require that you prevent the nose from rising to an excessive pitch through the use of forward pressure on the elevator, you will not damage the controls. As you add full power, be prepared for any nose-up tendencies, then retrim the airplane to neutralize the elevator control inputs needed to maintain the correct climb speeds.

Approach Controls

During the approach you will use the flight controls, the rudder, ailerons, and elevator in concert with each other. We have already discussed the basics of flying the airplane in the pattern, to final approach, the flare, and finally touchdown and rollout. During a normal landing you will use ailerons to maintain the airplane in a wings-level attitude, making your aileron inputs smooth and small, just enough to compensate for any small changes. During your turns in the pattern you will need to keep the ailerons and rudder coordinated, which we have covered in great detail. One thing to be aware of during the approach is that it may be necessary to input a small amount of left rudder during the descent to compensate for any right-turning tendency due to the nose-low attitude. Depending on the plane, the amount of left rudder may range from very little to a noticeable amount.

Once you are established on final, keep the intended landing zone fixed in its position on the windscreen, and scan the runway for any other traffic. Also scan taxiways and intersecting runways for traffic. While you may know what the active runway is, not every pilot listens to the radio or pays attention to other traffic in the pattern. While I worked at an airport, a number of years ago two airplanes collided at the intersection of two runways, killing both pilots. One was taking off on one runway, while another was landing on an intersecting runway. I'll never forget the dark plume of smoke that rose from the wreckage that day, or the lesson we can all learn from accidents like this. Never assume that traffic is aware of you. Make sure you are watching out for other planes, and look in places where it might seem unlikely they will come from. Those are the ones that can really surprise you.

As the plane approaches the runway and you begin the flare, use the ailerons to maintain a wings-level attitude, setting the plane up to touchdown in a nose-high attitude. Avoid pitching the nose up too much or too quickly, as this can result in the plane ballooning back up into the air or stalling while you are still well above the runway. As the nose does pitch up, you may need to input a small amount of right rudder, once again to compensate for any left-turning tendency that may present itself in this nose-high attitude. In no-wind conditions, the longitudinal axis of the plane will line up with the runway centerline, and no crab or slip will be needed to keep the airplane centered on the runway. After the plane touches down, hold the nosewheel off the runway, gently lowering it to the ground with the elevator. Use rudder and/or nosewheel steering to keep the plane tracking straight down the runway until you slow and turn off.

Don't rush the use of controls or overcontrol the plane during the approach. One of the most common errors I have seen students make as they are learning normal landings is dropping the nosewheel to the runway very quickly. Just as leaving the nosewheel on the runway too long during takeoff can cause damage or control problems, so can dropping it to the runway too quickly after landing. Unless conditions warrant getting the nosewheel down, keep it off until the plane slows, then slowly lower it to the runway. Keep all control inputs smooth and gentle, anticipating how the plane will react. With practice, you will find you can consistently flare, touch down, and roll out in a smooth, controlled manner. Critique yourself during each takeoff and landing, and look for ways to improve how you fly. You can be your best teacher if you are aware of what is going on. Now let's take a look at short-field takeoffs and landings.

SHORT-FIELD TAKEOFFS AND LANDINGS

In this section we will discuss short-field takeoff and landings, one of my favorite landing scenarios because it puts together so many of the elements of flight that we have discussed so far in this chapter, including pattern work, airspeed control, and touchdown zone accuracy. There's nothing like the challenge of setting up to land on a short runway, picking the touchdown spot near the runway threshold, and flying the approach right down to the exact target you had on the runway. Pilots who are not confident in their ability to get into a short field, even if it falls within the minimum runway lengths recommended in the pilot's operating handbook, are often inconsistent in one or more of these areas. As you read through this section, keep in mind that you must be proficient at all the elements of takeoff and landing to smoothly and efficiently execute them every time.

Chapter Six

Short-Field Takeoff

The short-field takeoff is used when you are flying out of a runway that has a length that is near the minimum needed for your aircraft at the gross weight and weather conditions for that takeoff. You will need to consult the pilot's operating handbook and use the appropriate charts to determine the minimum runway lengths needed for your airplane, and that will take into account the gross weight of the plane, wind conditions, density altitude, flaps settings, runway conditions, and other factors affecting performance. You may also need to consult the handbook for takeoff distances if you need to clear the ever-present 50-foot obstacle so thoughtfully placed at the end of the runway. Most operations manuals also cover the distance needed to clear an obstacle of that height. In either case, the idea is to get the plane off the ground using the minimum runway and climb at the appropriate airspeed to clear any obstacles. Let's step through an example of a short-field takeoff to understand the procedures. For the purposes of our example, we will assume you have completed the preflight and pretakeoff checklists, cleared the pattern, and received permission for takeoff. Trim will be set to the takeoff position, and flaps must be set to the amount directed by the operations manual for the plane you are flying. Many manufacturers recommend 10 degrees of flaps for a short-field takeoff to help the plane get off the ground faster and climb better, but you must always consult the manual to get the correct flap setting. Don't feel that if some flaps are good, more are better. If you exceed the recommended flap setting, the greater drag caused by the additional flaps may dramatically increase the distance of your takeoff; in tight situations it may even prevent you from being able to lift off before you run out of runway.

In addition to a recommended flap setting, the manufacturer will have a short-field takeoff speed. If you attempt to lift off before you reach that airspeed, you may stall and settle back to the runway, but if you let the airspeed accelerate past that airspeed, you will also use more runway. Know the correct rotation speed for short-field takeoffs. You will also need to know what climb airspeed the manufacturer wants you to use for short-field takeoffs. This is often V_x, which allows you to achieve the greatest altitude gain over a given distance, an important consideration if you need to clear a tree or building near the end of the runway.

When executing a short-field takeoff, you will want to use all available runway, so it may be necessary to back taxi to the end of the runway in order to have all available runway accessible during the takeoff. Once you are aligned with the runway, apply full throttle in a smooth, controlled manner, noting the engine gauges to be sure you are in the green on oil and fuel pressure. As you reach the recommended airspeed, raise the nosewheel from the runway. Hold the elevator in a neutral to slightly tail-low position as the plane accelerates. As you reach the recommended rotation speed, smoothly feed in up-elevator to raise the nose of the plane to the correct climb attitude. With practice you will be able to anticipate the correct attitude and be able to smoothly position the nose at that attitude. Your climb speed will depend on whether you need to clear an obstacle or can transition to a cruise climb airspeed, but in either case know what the value is before you begin the takeoff roll.

Your takeoff will be at a slightly slower airspeed than a normal takeoff, resulting in reduced flight control effectiveness until the plane accelerates. This means you may need more right rudder to hold the ball in the center during the climb and will need to use more elevator and aileron to hold the correct pitch or maintain a wings-level attitude. Don't overcontrol because of the reduced control effectiveness. Instead, anticipate what the plane will do and compensate for the reduced control authority. When you are executing a short-field takeoff on gusty or windy days, you may need to let the plane accelerate to a higher airspeed before rotation to assure that you have sufficient control authority to control the airplane in those conditions. As a result, you may also need to allow for greater runway lengths to accommodate the extra airspeed you will be using. Keep in mind that if the conditions are very windy, or hot, or anything else that can affect the performance of the plane, one option is to delay your takeoff until the conditions have improved to allow for a safe takeoff.

Some common errors committed during short-field takeoffs include not using the full runway length for takeoff, an incorrect pitch attitude, and incorrect climb airspeeds. Through practice you will be able to eliminate these and other errors, becoming proficient in short-field takeoffs. A short-field takeoff requires you to fly the plane near the limits of its flight envelope. A mistake in technique or judgment could result in serious damage to the plane or injury to you or your passengers.

Short-Field Landing

Short-field landings are one of my favorite landing situations. Like the short-field takeoff, you are flying the airplane at its limits, landing it in as short a distance as possible while still maintaining control of the plane and keeping safety as the highest priority. Consistently and accurately landing the plane at a given spot on the runway and getting stopped in the minimum distance requires that you fly the airplane at exact airspeeds with a consistent glideslope and touch down on your intended landing spot. A short-field landing utilizes all of the skills you develop as a pilot and is a good measure of how well an individual understands the interrelationships of all of these skills.

In this section we will discuss the how-to's of short-field landings, walking through an example. In concept it is very similar to a normal landing, differing in the fact that you are flying at a slower airspeed and may not have as great a margin for your touchdown point. It is not uncommon for pilots who are first making short-field landings at a runway that is actually very short and narrow to feel a certain amount of apprehension as they look down at the runway after the turn to final. If the runway is bordered by trees or other obstacles, the runway can appear that much smaller. Use the operating handbook to determine the minimum runway lengths you will need for the plane you fly. Density altitude, runway surface, temperature and winds are just a few of the factors that affect how much runway will be necessary to get your plane safely down and stopped. Make sure you understand how to use the tables for finding landing distances; they can differ for each model aircraft or aircraft from different manufacturers. Now let's walk through an example.

Flaps Settings

Flaps are an important part of short-field landings. As we have already established, flaps help reduce the stall speed and allow a steeper rate of descent without an increase in the airspeed. This steeper glideslope will allow you to clear those towering pine trees located at the end of every short runway and not end up as far down the runway before your wheels touch down. Check the operations manual for the plane you are flying for the correct flap settings for a short-field landing, but very often manufacturers recommend the use of full flaps once you are on final approach. In cases where I am using full flaps, I like to put one notch on after I have turned to the base leg, the second notch of flaps on after the turn to final, and extend the third notch on short final; then I know I have the runway made. This can be adapted as necessary to the airplane and conditions you are flying under.

As you apply each notch of flaps, it is important to maintain the correct airspeed to hold a constant glideslope and relation to the touchdown point. This becomes even more important in short-field situations, where floating down the runway because of excessive airspeed can use up runway and result in a go-around. Retrim the plane as you extend each notch of flaps to reduce your workload. You should also become familiar with the feel of the plane as the airspeed is reduced to the value recommended by the manufacturer for short-field landings. It is likely you will be flying 10 knots or so above V_{so} while on short final, and the controls will become sluggish and soft. Anticipate the need to correct for attitude and airspeed, locking the airspeed indicator at the correct value. Listen to the sound of the slipstream and engine for cues as to the speed of the plane. These sensory inputs can help you learn to divide your attention between looking outside the plane to monitor your progress through the approach and knowing how fast the plane is traveling. After a little practice, it will become almost second nature to be able to "sense" the speed of the plane.

Trim Settings

Use trim in the same manner we have discussed for previous landing scenarios. By reducing the pressures you must maintain on the control yoke, you can reduce your workload and be able to focus on other aspects of the approach. In setting up for a short-field landing, using trim to your advantage can help you lock in the airspeed and glideslope and make the task that much easier. Reset the trim after each notch of flaps is extended, setting it to hold the airspeed at the correct value. You will find that with a little practice you can anticipate the amount of trim you will need to hold the correct airspeed.

Flare, Landing, and Rollout

The short-field approach can have a steeper glideslope than other landings, depending on the runway length and airspeeds used. Figure 6-12 shows a short-field approach using power on during the approach until an obstacle at the end of the runway is cleared, then power off to help increase the angle of the glideslope. This type of approach can help you hold a shallower glideslope until the obstacle is cleared. A drawback is that as power is reduced, the glideslope and rate of descent increase. You will need to anticipate the effect this will have on your flare; the higher your rate of descent, the more rapid the flare will need to be to arrest your descent for a smooth landing on the runway. Figure 6-13 shows a different glideslope and approach, this one using a more constant power setting and glideslope through the entire final approach. In this approach the plane is on a more constant glideslope, and it can be somewhat easier to judge where you are going to actually touch down on the runway because of the fixed rate of descent. This approach requires a better eye at judging your clearance over an obstacle, but with practice can be an alternative you may want to consider.

At this point let's assume you have set up airspeed at the value recommended by the manufacturer for short-field landing. You have full flaps extended and are on the correct glideslope with the intended touchdown point locked in the windscreen. Your airspeed is likely slower than a normal approach and the controls are somewhat soft compared to approaches flown at higher airspeeds. Keeping the control inputs coordinated is extremely important at this point because of the reduced airspeed. You do not want to set yourself up for a stall/spin at this point because of incorrect use of flight controls. The other thing that becomes more important is not hanging the airplane on the prop, using

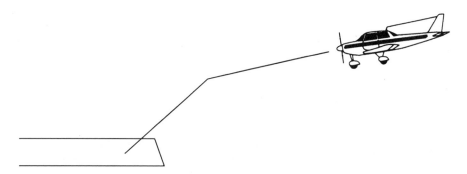

Fig. 6-12 *Short field approach: power on/off.*

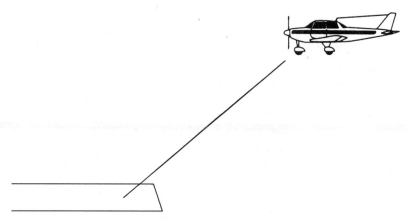

Fig. 6-13 *Short field approach: constant power settings.*

engine power to drag the airplane in over obstacles or to the runway itself. This can be a very tempting option as pilots try making a real short-field landing at a very short runway. An approach using engine power in this manner sets you up for problems if the engine fails, a topic we discussed previously that becomes even more important now.

Even with a well-flown glideslope, the rate of descent will be higher during a short-field approach. You will probably need to begin the flare a little higher to give you time to arrest the rate of descent and put the airplane in the correct attitude for touchdown. Because of the slower airspeeds you are flying at, more elevator input may be necessary during the flare as well. The flare should put you in a nose-high attitude, the main gear touching down with the nosewheel held off the runway. In some cases you may find that the landing can be a little harder than a normal approach, but not significantly so. The idea is to touch down as gently as possible, not drive the airplane onto the runway. To avoid damage, control lowering the nosewheel to the ground. Don't be tempted to slam it to the runway in an attempt to get stopped sooner.

Maximum braking may be necessary when you are landing on very short runways, so it may be necessary to retract your flaps after the nosewheel touches down to put more weight on the main gear. During breaking you may need to hold nose-up elevator to prevent the plane from pitching down and placing excessive forces on the nose gear, possibly digging it into a soft runway surface. You will need to balance the amount of up elevator between too much, which could raise the nose again, and too little, which could damage the nose gear. Feel the airplane during the landing and adjust your control inputs accordingly.

Several common errors made during short-field landings include poor airspeed control, not flaring at the correct altitude, and not using enough elevator backpressure to get the plane to the correct landing attitude. Other errors include improper braking and letting the nose gear slam onto the runway after the mains touch. Short-field landings are a great deal of fun and can build confidence in pilots who are proficient at them. These landings use so many of the skills we need as safe, competent pilots that you should practice them whenever you can.

SOFT-FIELD TAKEOFFS AND LANDINGS

Not every runway is sixty feet wide, ten thousand feet long, and paved in concrete. For those occasions when you must fly into or out of a grass runway, or a snow covered runway, you will need to know the best way to take off or land on those soft runway surfaces. Soft runways can increase the distance needed to accelerate to flying speed due to the drag those surfaces create as the aircraft's wheels roll over them. In some cases the drag created by soft snow, tall grass, or soft ground can prevent a plane from being able to reach flying speeds all together. In this section we will discuss techniques that can help reduce the distance it takes to get the airplane up to flying speed and off the runway in as safe a manner as possible. Landing on soft runways can also pose unique challenges to you. Landing at too high an airspeed can cause the plane to dig into the surface while still traveling too fast, causing potential handling problems. Landing in the improper attitude can also cause the nosewheel to dig in while the plane is traveling too quickly and potentially flip the plane over on its back. We will also cover soft-field landing techniques that can help reduce the chance of improper landings on soft-field surfaces.

Soft-Field Takeoffs

We will now look at the proper techniques for a soft-field takeoff. You will see that this is very much a finesse maneuver, requiring that you know the airplane, feeling what it is telling you as you accelerate along the runway. You will also need to know when to abort a soft-field takeoff, or when not to attempt a takeoff at all. Not all runways are fit for takeoff. Deep snow, muddy runways, or tall grass can prevent you from taking off at all. If you are unable to get to flying speeds, you need to know exactly where you will abort the takeoff and allow yourself enough distance to safely come to a stop. Before any soft-field takeoff, you should inspect the runway, noting spots along it that may cause trouble or prevent the takeoff. Grass runways may be unusable in the spring or after a heavy rain, the surface becoming so soft that you will not be able to accelerate to flying speed.

Before we look at soft-field takeoff techniques, let's briefly review a concept that is extremely important to this aspect of flying—ground effect. We already know that ground effect can reduce drag when we are close to the ground and diminish the angle of attack needed to produce a given coefficient of lift. This can help us get off the ground before we could without the benefit of ground effect. This is very important during soft-field takeoffs, where we want to reduce the drag caused by the runway's surface by getting off the ground as quickly as possible. Through understanding ground effect and the fact that if we leave the runway and remain relatively close to it, we know we can reduce the braking effect caused by a soft runway surface. But the danger of using the benefits of ground effect is not knowing the limits, of pulling up too far from the runway before adequate flying speed has been achieved. In these cases the plane can settle back to the runway or stall and not have sufficient altitude for recovery. When you are executing a soft-field takeoff, you want to remain close to the runway after liftoff until you have accelerated to safe flying speeds. This is very much a part of the finesse you will need to consistently, safely execute a soft-field takeoff. As we discuss the soft-field takeoff techniques, keep in mind that you must balance between clearing the snow, grass, or mud that slows the acceleration, and attempting to fly away from ground effect before the plane is able to do so.

Flap Settings

Several factors come into play when considering flap settings for soft-field takeoffs. Manufacturers will indicate the correct flap setting for the plane you fly in the operations manual. This normally is no greater than one notch of flaps, but may differ for the plane you fly. If you use a greater flap setting than the manufacturer recommends, the additional drag caused by the flaps may prevent you from being able to achieve flying speed in the runway distance available.

There may be cases when you do not want to use any flaps. If the runway is very soft and muddy, or covered with snow or water when temperatures are close to freezing, having flaps extended during takeoff may result in damage to the flaps during the takeoff run.

If mud is thrown up by the wheels, it can damage the flaps or get lodged in the flap actuation mechanisms and prevent their proper action. If water or snow is thrown up onto the flaps, it could freeze and prevent you from retracting the flaps after takeoff. When conditions may result in flap damage, it may be best to take off with the flaps retracted, but you should consult the operations manual for the plane before considering this option. Low-wing aircraft may be more susceptible to this situation than a high-wing airplane.

Assuming that conditions permit the use of flaps during a soft-field takeoff, you will want to maintain the flaps in the extended position until you have accelerated to safe flying speeds and climbed to a safe altitude before you retract them. If you are at too slow an airspeed when you retract the flaps, it may cause the plane to stall at too low an altitude to recover. Retracting the flaps may also cause the plane to settle back to the ground. When you are executing a short-field takeoff, the runway may no longer be under the plane, but behind it in this particular scenario. Touching down off runway could be a very serious end to your flight, so be certain you are several hundred feet above the ground before you retract flaps.

Trim Settings

Trim settings are very important during soft-field takeoffs. Because it is necessary to hold almost full aft elevator input during the initial takeoff run, some pilots are tempted to set the trim control to a greater nose-up attitude than is recommended by the manufacturer. If too much nose-up trim is set, the plane may have a tendency to pitch up and either stall or climb too rapidly out of ground effect. Use only the trim setting recommended in the operations manual for a soft-field takeoff.

Soft-field takeoff

We know what a soft-field takeoff is—getting the airplane off the runway surface as soon as possible to reduce the drag caused by the runway surface. In this section we will walk through a soft-field takeoff, discussing the control inputs and techniques commonly used. Before attempting a soft-field takeoff, you should inspect the runway to be certain it is safe to take off on. Potholes or deep muddy spots can catch a wheel during the takeoff run, resulting in damage to the airplane or loss of directional control. The runup should be done on a hard surface to avoid getting stuck on or near the runway's soft surface. Once the runup is completed, you will want to taxi onto the runway, holding the elevator in the aft position as you taxi to help keep the weight off the nose gear. This reduces the chances of damaging the nose gear or allowing it to become mired as you taxi. Turn onto the runway maintaining a safe speed; do not stop on the runway, but keep the plane moving as you turn to align with the runway centerline.

As you align with the centerline, keep the elevator control fully aft and apply full engine power. The nosewheel should rise from the runway surface as the plane accelerates; you want to hold it off the runway, but not at too high of a pitch angle. The idea is to prevent the plane's nosewheel from digging in as the plane accelerates, yet not have it at such an attitude that the plane pops up from the runway before it is able to fly and stalls. As the plane continues to gain speed, you will need to reduce the elevator backpressure in smooth, small increments. Be prepared for the left-turning tendencies and input enough right rudder to keep the airplane tracking straight down the runway centerline. You will begin to feel the plane getting light on the main gear, indicating that it is getting close to flying speed. At this point it becomes very necessary that you "feel" the airplane as it becomes ready to fly. As the plane leaves the ground, you want to reduce elevator backpressure enough to level the plane just above the runway and while it is still in ground effect. Now you should let the plane accelerate while still in ground effect until you have reached V_x or V_y, depending on the climb you want to establish.

The difficult part of soft-field takeoffs is knowing how much elevator input to use as the plane breaks ground. If you use too much nose-up elevator, the plane can stall; with too little elevator the plane can settle back to the runway. Practice is the best way to become familiar with the amount of elevator needed during a soft-field takeoff. Begin practicing on a long, hard surface runway until you become comfortable with the amount of elevator and rudder needed to execute the takeoff. A long runway also gives you room ahead in case you happen to settle back to the runway after the initial liftoff. It also lets you safely accelerate in ground effect with sufficient runway in front of you as you fly above the runway.

A number of errors are common to pilots who are learning soft-field takeoffs. Insufficient right rudder can cause the plane to veer to the left as you raise the nose. Pilots are often hesitant to input enough aft elevator to get the airplane's nose gear off the runway early in the takeoff run. This is not uncommon, given that they are taught not to establish too great an angle of attack during takeoff from their first takeoff lesson. Another common error is attempting to fly out of ground effect before the plane has achieved safe flying speeds. Again, most pilots are new to flying low over the runway at slow airspeeds as they learn soft-field takeoffs. You should also note that on hot days, or when the density altitude is very high, it may take a great deal more runway to execute a soft-field takeoff, so be certain to use the performance charts before you attempt to fly.

Soft-Field Landings

Soft-field landings are a lot of fun to execute, and in many ways resemble a short-field landing, but with an important difference. The intent of a short-field landing is to touch down as slowly as possible to reduce the amount of runway necessary to get down and stopped. A soft-field landing is also executed at minimum touchdown speed, but you are trying to avoid becoming mired in the soft surface of the runway. Combined with not getting stuck, you also do not want to end up nosing the airplane over due to the increased drag caused by the surface on the landing gear, as it can be rapidly slowed by the runway's braking action. While working as a line boy, I remember watching a Cessna 170, a taildragger, land in the early spring. The runway the pilot chose to land on was a sod strip, which was still very soft that early in the year. As the plane touched down, the main gear dug into the soft ground, and the plane nosed over, the propeller cutting swathes through the grass and mud. Fortunately, there was no propeller or engine damage because the ground was so soft, but for me that drove home how important it is to understand the correct procedures for soft-field landings.

When making soft-field landings, you should anticipate the possibility that the plane may have a tendency to nose over as you touch down. Single-engine airplanes may have landing speeds of fifty to one hundred miles an hour, or greater. The rapid deceleration can have a strong tendency to cause the plane to pitch nose down. While making soft-field landings, you must maintain a nose-high attitude during the landing to help reduce the chance that the plane will pitch down excessively. This combines the need to land at slow airspeeds with maintaining the proper landing attitude and holding it through the rollout.

To prepare for a soft-field landing, it may also be necessary to make a low pass over the runway if traffic permits to determine where any potential problem spots might be on it. Pooled water, muddy ground, or deep snow might be areas that you should avoid during a landing. In some cases it may be necessary to land off to one side of the centerline, or to land some distance down the runway to avoid problem areas. Asking for this information over unicom may also be a useful way to learn about runway conditions. At night the problem is further aggravated by lack of illumination of the surface, making it very difficult to see the runway at all. Whatever the situation, be prepared for the conditions and how they may affect your landing.

Flap Settings

Flaps will help reduce your landing speed and are therefore an important component of soft-field landings. Full flaps are recommended by many aircraft manufacturers for soft-field landings, but depending on runway conditions it may be wise to use a lesser flap setting to prevent debris from being thrown up and damaging the flaps or their actuating mechanisms. This can be more of a factor with low-wing airplanes than with high-wing, due to the closer proximity of the wheels to the wings. The type of surface, and how soft it is, can also be a factor. Mud or gravel could be more of a problem than grass or water, but each case is unique and must be evaluated. As you make a soft-field landing, keep in mind the conditions and the recommendations of the aircraft's manufacturer when picking the flap settings.

Trim Settings

Trim settings during soft-field landings should follow those listed in the operations handbook for the plane you fly. Normally you will trim the plane for the airspeeds you want to maintain during the approach, just as with any landing. The airspeeds will likely be at the slower end of the airspeed range during the final approach leg to accommodate the need to touch down at minimum airspeeds. Once again, if you set the trim correctly during the approach, you can reduce the physical work load and allow yourself the ability to focus on other aspects of the landing. Not having to work as hard to get the nose to the proper attitude as the main gear touches down on the runway can be a benefit of using correct trim settings. This can be especially useful for those aircraft that tend to be more nose heavy and require a fair amount of backpressure on the control yoke to hold the nose gear off the runway's surface.

Soft-Field Landing

We've already alluded to the fact that a soft-field landing is very much like a short-field landing. The main difference is that you want to hold the nose gear off the runway during the landing and keep it clear of the ground as long as possible after touchdown. The runway's surface may cause the nosewheel to dig in, and reducing the weight on it can help prevent a possible mishap. Once the plane has slowed, you will also want to avoid getting stuck on the runway, depending on the runway's makeup. It would be very frustrating to make a smooth, safe landing on a soft, grass runway only to become stuck in mud after the plane slows. In this section we will look at the techniques you should use to make a safe, controlled landing on a soft runway and still be able to taxi to the ramp. Keep in mind that procedures may need to be modified depending on the runway's surface, or the plane you are flying and its limitations.

We will assume that the approach has been set up and flown just as with the previous examples we have discussed, with the final approach speed at the value recommended by the aircraft's manufacturer. In this example we will also assume that full flaps are extended and that no damage will be caused by the runway's surface. Further, trim is set to hold the elevator pressures neutral at the final approach speed. Our flare will be at the normal height, with an understanding that the slower airspeeds may result in the need for additional elevator input to achieve the proper pitch attitudes. With a soft-field landing, we want to really work at flaring at the correct altitude and setting the main gear lightly on the runway with the nosewheel well off the ground. Dropping the airplane to the runway after flaring too high, or driving the plane into the ground after flaring too low, could result in a greater rate of sink than a smooth, properly executed flare. The outcome of either of these situations could be that the plane becomes stuck very quickly, so being able to accurately judge your height during the flare is very useful.

As the plane touches down on the runway, hold the elevator control in the full aft position as the plane slows to hold the nose gear off the runway. Gently lower the nosewheel as the plane slows, avoiding any tendency to drop it to avoid letting it dig into the soft surface. Once the plane is on the runway, it may decelerate very quickly due to the braking action of the soft surface. At that point it may be necessary to add a slight amount of engine power to keep your forward momentum up, and to increase the amount of airflow over the elevator. This additional airflow can give you just a little more control authority and allow you to maintain a lighter load on the nose gear.

Be prepared for possible directional control problems if the runway contains ruts, potholes, or any other obstacle that one of the landing gear could dig into. Do not overreact if one of the landing gear does cause the airplane to swerve to some degree. Make your corrections as smoothly and with as little control input as possible. If you need to add power to help keep the plane moving forward, use small additions there as well. You don't want to add so much power that it pops the nose gear off the runway or puts too much stress on the main gear as the plane tries to accelerate on a very soft runway.

Common errors for soft-field landings include touching down at too high an airspeed, judging the flare height incorrectly and causing the main gear to dig in to the runway, and not keeping the elevator control back after the plane has touched down. With practice, soft-field landings become as easy to execute as any other landing. The real test comes the first time you practice on a truly soft field, like a grass runway.

What you might find is that if you put the landing elements together just right, you'll get a softer touchdown than you have ever experienced—not a bad ego booster for any pilot.

CROSSWIND TAKEOFFS AND LANDINGS

Our discussions so far in this chapter have centered on a number of takeoff and landing techniques that meet specific situations such as short runways or soft runways. And we have also reviewed the need to be aware of how to compensate for wind as you fly ground tracks in the pattern. But you have probably already asked yourself the questions, "How do I correct for the effects of wind during the takeoffs and landings themselves?" Having the airplane crabbed into the wind at touchdown on the runway can cause the plane to execute some interesting changes of direction since the wheels will not be parallel to the direction the plane is moving. To avoid this not so little problem, and others associated with the effects of wind during takeoffs and landings, we must learn how to keep the longitudinal axis off the plane parallel with the runway while preventing the plane from drifting to one side or the other.

Wind can have a tendency to lift the wings of the airplane not only during the takeoff and landing rolls, but also while you are taxiing the plane. As you can see, you need to have an understanding of what the effects of wind can be during all phases of flight, from the time the engine starts until you shut it down. You also need to be able to compensate for the wind while holding the plane in the proper attitude and maintaining safe control of the aircraft. In this section of the chapter, we will cover a number of different aspects of crosswind takeoffs and landings. Let's begin with crosswind taxiing and takeoffs.

Crosswind Taxiing

During the process of taxiing the plane out to the runway, or back to the ramp, you need to be able to maintain safe control of the plane during windy conditions. Figure 6-14 shows the different directions of the wind in relation to the aircraft and the control inputs that are necessary to reduce the effects the wind has on the plane. As you taxi the plane under windy conditions, you will find that it will have a tendency to weathervane into the wind as a result of the rudder and vertical stabilizer. Much like a weathervane on top of a barn pivots and points into the wind, your airplane will attempt to point its nose into the wind. Additionally, wind has a tendency to get up under a wing and can actually lift the wing and landing gear. In severe conditions this can cause the plane to flip up on a wingtip, or onto its back. Either case can result in significant damage to the plane or injury to the occupants. Correct use of the controls can minimize the ability of the wind to cause aircraft control problems.

As you look at Figure 6-14, note that it indicates the correct control inputs for aileron and elevator for each quadrant the wind is blowing from in relation to the plane. If the wind is coming from the left forward quadrant, up aileron on the left-hand wind and neutral elevator is in order to prevent the left wing from rising. Neutral elevator will help keep the nose on the ground. If the wind is from the right forward quarter of the airplane, you

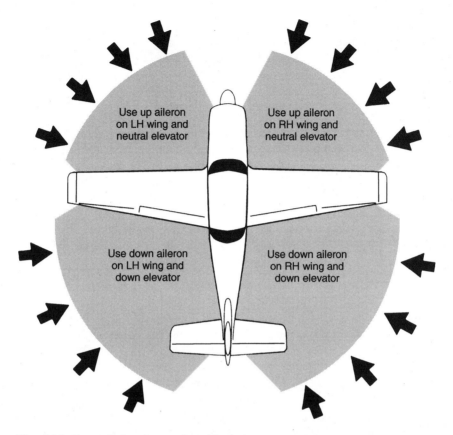

Fig. 6-14 *Crosswind taxi control positioning.*

will want to use up aileron on the right-hand wing and neutral elevator once again. As with the left wing, the raised aileron helps the wind force the wing down, instead of up.

When the wind is from the left rear quarter of the plane, you need to lower the aileron on the left wing and use nose-down elevator. In this situation the lowered aileron will help the wind to flow over the top of the wing, making it easier to keep down on the ground. The down elevator makes it more difficult for wind to get under the elevator and horizontal stabilizer. If strong winds do manage to get under the tail empennage, this can raise the tail of the plane and force the nose down. The danger is that the propeller can dig into the ground, or the plane can flip over on its back. With a right quartering tailwind, you will need to use down aileron on the right-hand wing and use down elevator, the reasons being same as with the left quartering tailwind. These control inputs work best for nosewheel aircraft and will need to be modified for tailwheel planes. Check the manufacturer's operations manual for further information for taildraggers.

As a final note on taxiing in windy conditions, if things are too severe you should always be willing to consider postponing your flight until the wind has subsided. I have worked at airports where it was necessary at times to have line personnel hold the wing tips of a plane as it taxied back in after landing. The flight controls provided enough compensation for the wind while the plane had sufficient airspeed, but once slowed to taxiing speeds the plane was at the mercy of the wind. Don't let your desire to complete a flight cloud your judgment.

Crosswind Takeoffs

Crosswind takeoff techniques apply to each of the takeoff scenarios we have covered so far in this chapter. As you read this section, keep in mind that you will need to blend in the crosswind control inputs to correct for the wind while executing the takeoff of choice. For instance, if you are executing a short-field takeoff with a left quartering headwind you will need to combine the short-field takeoff steps with left aileron and potentially left rudder during the ground roll to keep the plane under control.

Depending on the strength of the winds, you may need to consider letting the airplane accelerate to a higher airspeed before rotation to the takeoff attitude. This additional airspeed will provide greater control authority, and this may be very necessary, depending on the strength and direction of the wind. We should normally take off and land into the wind; these headwinds help reduce the amount of runway needed. If wind speeds are significant, the reduction in runway lengths can be significant, so use them to your advantage. However, there may be cases where it is necessary to make a downwind takeoff or landing. For example, if a tall bluff or grove of trees is present at one end of the runway, and you are concerned about actually being able to clear it using a standard into-the-wind takeoff, it may be wiser to execute a downwind takeoff. If you consider this option, you will need to calculate the additional runway needed to take off or land. A tailwind during either operation can result in a large increase in the amount of runway needed, so be certain you know the distances before you make a decision.

Flap Settings

Like soft-field operations, crosswind flying can affect your decision to use flaps. The additional wing area created by flap extension gives wind more surface area to affect the plane. The lower airspeeds that flaps let us take off or land at also result in reduced control effectiveness, making it more difficult to compensate for the wind. Most aircraft manufacturers only recommend the use of one notch of flaps during takeoff, but you must consider what effect this will have on the plane during a crosswind takeoff. Experience will help you gauge the correct amount of flaps to use during both takeoff and landings, and the manufacturer may also have additional tips for crosswind flap settings. Whatever the conditions, be prepared for the effects wind may have on the plane and how flaps can play a contributing factor in your ability to control the airplane.

Ground Roll and Rotation

During a crosswind takeoff, you will need to apply control inputs that keep the wings from being picked up by the wind, keep the plane from weathervaning into the wind, and allow you to hold the longitudinal axis of the plane parallel to the runway centerline. In our example we will assume a left quartering headwind during the takeoff. A normal takeoff will be executed with crosswind control inputs applied as necessary.

After you taxi onto the runway and align with the centerline, you will want to roll in full left aileron. This will raise the aileron on the left wing, helping to force the wing down as wind flows across it. Normal trim and neutral elevator will be used at the beginning of the takeoff roll. Apply full throttle and release the brakes, maintaining the full left aileron input as the plane accelerates. Elevator use will be the same as for any normal takeoff. As the airspeed increases and additional control authority is established, you will want to use just enough left aileron to prevent the upwind wing, the left wing in this case, from rising due to the crosswind. If too much left aileron is left in as the plane accelerates, it can actually cause the right wing to rise, a situation you also want to avoid. Roll out just enough aileron to keep the wings level during the takeoff roll as you continue to gain speed.

It is very likely that you will need to input additional right rudder during our takeoff example to prevent the nose of the plane from weathervaning to the left. The amount of rudder necessary will also become smaller as the airspeed increases. Crosswind takeoffs and landings can cause additional tire wear as you use nosewheel steering to maintain the correct direction during the takeoff roll, and you want to avoid overcontrolling during the ground roll. As you rotate the nose to the takeoff position, it may be necessary to add additional right rudder to hold the nose straight down the runway. As the nosewheel leaves the ground, the directional control it provided will be lost. Additional right rudder can be input to compensate for this loss of nosewheel steering. If winds are very high, or your airspeed is slow at rotation time, there may not be sufficient rudder authority to keep the aircraft from weathervaning. Letting the plane accelerate to a slightly higher airspeed before rotation can help prevent this problem. Keep enough left aileron in to hold the wings level as the nose rotates to the takeoff attitude.

When the plane breaks ground, you will want to establish a crab into the wind that will prevent the crosswind from pushing the plane off the ground track of the runway centerline. This can be done by turning the plane slightly into the wind, just as with our ground reference maneuvers. We are now flying and can use standard ground reference maneuver techniques to hold the proper attitude and ground track as we continue to climb.

Common mistakes made during crosswind takeoffs include not rolling in aileron, or rolling in insufficient aileron to compensate for the wind. Always use full aileron input as you begin the takeoff run and reduce it as necessary. The other common mistake is not to anticipate the need for rudder to overcome any weathervaning tendencies, which allows the airplane to drift across the runway. This is very important when flying a retractable-gear airplane. These aircraft are sensitive to side loads on the landing gear, and if the sideward forces placed on the landing gear during takeoff or landing is too great, it could cause the gear to collapse. Now let's move on to crosswind landings, one of my favorite landing situations.

Crosswind Landings

Crosswind landings are a great deal of fun if you know how to execute them safely. Every airplane has limits for crosswinds that it can safely land in, but knowing how to safely land in crosswind conditions can make you a more confident pilot. In this section we will cover the major aspects of crosswind landings, from the crab you establish on final approach to the slip you use during the flare to touchdown. Properly done, crosswind landings can be accomplished as easily as any other landing. In fact, like the crosswind takeoff, you will find that crosswind landings must also be blended into the other landing scenarios we have already discussed in this chapter.

We will discuss a typical crosswind landing as our example in this section, but keep in mind that your airplane may require different techniques depending on the recommendations of the manufacturer. For the purposes of our discussion, we will assume the pattern has been flowing using wind correction techniques that were discussed in the ground reference maneuver chapter. We will also assume that the landing checklist has been completed and the plane is in the landing configuration for landing gear, carb heat, and other items. The landing itself will be a standard landing using crosswind techniques, but these techniques also apply to other landing scenarios such as short or soft field.

As you set up for a crosswind landing, you will want to consider how wind speed affects your ability to control the aircraft. Just as with the crosswind takeoff, control authority is very important in windy or gusty conditions. The additional airspeed also provides a higher margin to guard against an accidental stall of the airplane during the approach. For this reason it may be wise to carry additional airspeed during crosswind landing conditions. The amount of additional airspeed will vary with the strength of the wind. The stronger the wind conditions, the greater the need for added airspeed. You should use the airspeed ranges specified by the aircraft's manufacturer, and never exceed safe airspeeds. The additional airspeed will result in additional runway length requirements, so you will also need to weigh the airspeed against runway lengths available at the airport you are landing at.

Flaps and engine power during a crosswind landing should be adjusted to meet the conditions during your approach. Full flaps can help increase the rate of descent, decrease the stall speed, and allow you to fly more slowly. The trade-off is that full flaps also give the wind more wing area to affect and can increase controllability problems as you land. Maintaining a slight amount of engine power can also help improve your ability to deal with the wind by keeping the airspeed from getting too slow and losing control authority as a result. If you do elect to use additional power while on final approach, make sure you reduce the power to idle before touchdown to prevent the plane from floating down the runway.

Let's begin our approach on final with only a notch of flaps extended and the engine carrying about 1200 r.p.m. As you turn onto final, you will establish a crab into the wind to allow you to fly the centerline of the runway. This crab is the same maneuver we discussed during our ground reference maneuver chapter. With the airplane crabbed into the wind, you will watch the position of the target landing zone in the windscreen to determine if you are on the correct glideslope. Corrections to the glideslope are just as with any other landing. Trim is used to help maintain the correct airspeed and reduce the pilot's workload during the approach. As you first turn onto the final approach leg, you may find it difficult to accurately judge your drift from the runway centerline, but as you get closer to the runway it will become easier to determine the effects of the wind.

We cannot land the plane with it in the crabbed configuration. Like the crosswind takeoff, the longitudinal axis of the plane must be parallel to the centerline of the runway. Since the crab does not put us in this situation, we must somehow set up the airplane so that the longitudinal axis of the plane is correct and the plane is not drifting across the runway putting side loads on the landing gear. To prevent the problems that would be present when landing in the crabbed position, it is necessary to transition to the side slip, depicted in Figure 6-15. Also known as the wing-low method, this allows you to touch down with the plane aligned with the runway and not allow it to drift due to the effects of the wind.

As you begin to flare, the procedure is to lower the upwind wing into the wind while using opposite rudder to hold the longitudinal axis of the plane parallel to the runway. The amount the upwind wing must be lowered to counteract the crosswind is dependent on the speed and angle of the wind and the type of plane you are flying. Figure 6-16 shows a head-on view of the slide slip as the plane approaches the runway and touches down. As you can see, the plane touches down in a nose-high attitude, but because the upwind wind is lowered, the upwind main landing gear touches down first. The nose is held off, and the downwind main gear is lower to the runway in a controlled manner. Then the nose gear is also lowered to the runway. Rudder and nosewheel steering are used to keep the nose of the plane tracking straight down the runway, while aileron is rolled into the wind to prevent the upwind wing from being lifted by the wind. As you practice crosswind landings, it will become easier to gauge how much you need to lower the upwind wing and how much opposite rudder is necessary to hold the plane parallel to the direction of movement. The more you lower the upwind wing, the greater the amount of opposite rudder that will be needed.

Every airplane has a demonstrated crosswind component that the aircraft manufacturer documents through testing. You should never attempt to land an airplane when the crosswinds for a runway exceed the aircraft's maximum recommended value. Until you are proficient at crosswind landings, you should also avoid conditions that approach the maximum values for the aircraft. After touchdown you want to use enough aileron to prevent the upwind wing from rising, but not so much that you force the upwind wind into the runway surface. Crosswind landings require finesse, anticipation, and an understanding of the plane you are flying. A very real consideration is the fact that if you lower the upwind wing too much you can cause it to strike the runway during landing. This can result in anything from a damaged wing tip to a cartwheeling airplane, both dangerous and expensive propositions.

As you first begin to learn crosswind landings, it may be beneficial to fly a longer final approach and set up the side slip when you get to be about one-quarter to one-eighth mile from the touchdown point. This lets you observe the drift due to the crosswind and adjust the amount of wing lowering you will need to do to correct for winds. I like to work students into executing the side slip during the flare as quickly as possible as opposed to extended use of the side slip on final to avoid flying in a cross-controlled situation. With the ailerons rolling the plane in one direction, and the rudder forcing the nose in the opposite, we have a classic stall/spin situation if the plane should happen to stall

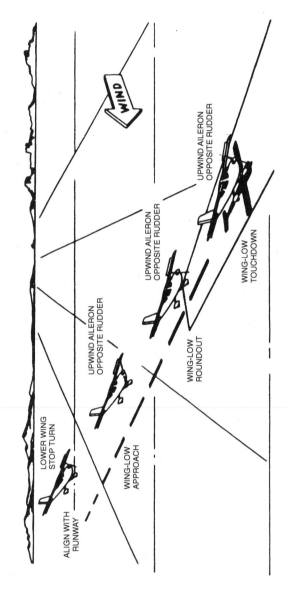

Fig. 6-15 *Crosswind landing techniques.*

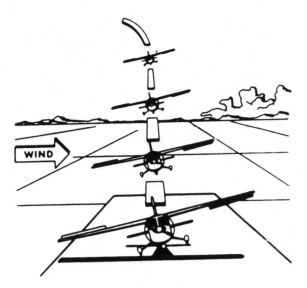

Fig. 6-16 *Crosswind landing techniques: front view.*

during the approach. Use of the crab on final avoids this, and transitioning to the slip during the flare reduces the chances of this type of mishap.

There are a number of common errors made during crosswind landings. These include not enough or too much upwind wind correct. As a result of using the incorrect amount of aileron input, the plane can drift in one direction or the other and introduce side loads on the landing gear during touchdown. In a similar manner, students frequently do not use enough rudder opposite the lowered aileron. This causes the plane to touch down not parallel to the direction of movement. When this angle to the runway's centerline is relatively small, this results in a hard or jerky landing, but in severe cases it can cause damage to the landing gear or loss of control of the plane. Some students are also uncomfortable with initially landing on just one main gear, feeling this may be unsafe. This method is very safe if properly executed and provides a great deal of control of the aircraft during the landing and rollout. If you work hard to master crosswind landings, you will find there is a great deal of satisfaction in being able to touch down with featherlike manner during a windy, blustery day.

ABORT PROCEDURES

As pilots we must be prepared for situations that can arise during both takeoff and landing. We look for other aircraft in the pattern, on intersecting runways and taxiways, and check the engine gauges during takeoff and landing to assure that all is in proper order as we execute a takeoff or landing. On those rare occasions when the unexpected takes place, we must be prepared to abort the takeoff or landing to avoid placing ourselves and our passengers in danger. In this section we will cover abort procedures for each of these situations. As you read this section, keep in mind that every situation is different and every airplane reacts differently. The methods for aborting takeoffs and landings in this section are guidelines and should be weighed against a given situation and the recommendations of the aircraft's manufacturer. The idea of this section is to present you with procedures that can help you avoid problems during an aborted takeoff or landing. As always, the first rule in any situation is to maintain control of the aircraft and keep thinking.

Takeoff Abort Procedures

There are any number of reasons to abort a takeoff: lower oil pressure or fuel flow, a rough-running engine, other traffic taxiing onto the runway. The list is almost endless. As you prepare for takeoff you should consider the length of the runway and how much you will need to stop if you do abort during the ground roll. Under ideal conditions, if you had not reached safe flying speeds by the point on the runway, you would abort the takeoff. The problem is that not all runways are long enough to allow us to do that. There will be those times, especially during short or soft-field takeoffs, that you may need most of the runway's length to accelerate to takeoff speeds. As you prepare to take off, keep safety as your primary concern and weigh the risks before you begin the takeoff roll.

If you do encounter a situation that results in an aborted takeoff, there are several things you need to keep in mind. First, maintain control of the aircraft. If pilots notice that oil pressure is low or another plane is taxiing onto the runway in front of them, they become so focused on the problem that they quit flying the aircraft. In any aborted takeoff you should be aware of the problem and keep part of your mind on controlling the airplane. After you notice a problem, immediately reduce engine power to idle and keep the plane tracking down the runway centerline. Depending on the plane, it may have a tendency to yaw in one direction or the other as power is reduced. Don't let this surprise you if your aircraft reacts in this manner. After power is reduced, use maximum braking to slow the plane as quickly as possible, but do not lock the brakes and cause the tires to skid on the runway's surface. *Maximum braking* is defined as the greatest braking pressure that can be applied without locking the brakes and resulting in a skid. If you do lock the brakes, it becomes more difficult to control the direction the airplane is traveling in and actually increases the stopping distance.

As you brake, it may be wise to hold nose-up elevator to help prevent the nosewheel from having too much weight shifted to it by the braking efforts. How much elevator will depend on the airspeed you are at, but you want to avoid raising the nose so much that the airplane leaves the runway. If you have flaps extended, it may be wise to retract them to help place additional weight on the landing gear, but don't become so focused on getting the flaps up that you don't fly the airplane. If impact is imminent, or a fire is present, you may also want to shut the fuel and magnetos off to help reduce the threat of fire. If oil pressure is low, shutting the engine down may also help reduce damage to the engine.

During my flying career I have had only one instructor actually perform a full aborted takeoff. As the plane rolled down the runway, he called for the abort and had me perform maximum braking, pull the mixture, and shut down the engine. This type of training can be hard on the airplane, so it may be difficult to find a flight school that will take the aborted takeoff to this level. To prepare for this situation as much as possible, though, you should practice with an instructor to the extent that they allow aborted takeoffs.

Landing Abort (Go-Around)

Aborted takeoffs can be pretty uncommon, but almost every pilot will execute an aborted landing at least once during his flying career. The most common reason I have executed a go-around is because of other traffic on the runway. Whether because another airplane has not seen me on final, or they waited too long to begin the takeoff roll, it has been necessary to abort the landing and re-enter the pattern for another try because someone else was using the runway I wanted to land on. For a lot of pilots I know, the first reaction to an aborted landing like this is anger at the other pilot who is blocking the runway. This reaction prevents them from readily focusing on flying the plane safely and can set pilots up for problems during the go-around. If you must execute a go-around, for whatever reason, don't let emotions get in the way of being a safe pilot. Like the aborted takeoff, the procedures in this section are guidelines. Use common sense, training, and aircraft procedures to fit a given go-around situation.

In addition to traffic on the runway, there are other reasons you may need to execute an aborted landing. But one of the hardest go-around scenarios to get pilots to admit to is aborting a landing because they are going to end up landing too far down the runway to safely stop. We can all easily accept the fact that we can't land and must execute a go-around if it is because of someone on the runway, or another plane landing on an intersecting runway. But it is more difficult to execute a go-around if it is because we have not set up to land at the correct spot on the runway. I have seen too many planes come in high on final approach, then shove the nose down as they point the plane at the originally intended landing spot, only to end up floating to nearly the far end of the runway because their airspeed is too high to land. After touching down, the pilots then stand on the brakes, trying to get stopped by the end of the runway. If you find yourself in this position, do not let pride get in the way of making good decisions. Execute a go-around and set up for the next landing. It's a lot easier to do that than to have to call the airport staff to pull your plane out of a culvert after the plane runs off the runway and into the grass, corn, or whatever else can occupy the ground around an airport.

Once you decide you need to execute a go-around, do not hesitate to make it happen. The biggest problem pilots encounter in an aborted landing is waiting so long to execute it that they have lost the window of opportunity to safely avoid a problem. Figure 6-17 shows the difference in altitude that can result from waiting, and how it can affect your ability to salvage the situation. After your decision is made, input full throttle to arrest the rate of descent. Ease the nose up to establish a positive rate of climb, then clean the airplane up after the proper airspeeds and rate of ascent have been established. During the landing you are at a relatively slow airspeed, so don't try to get the nose up too high or too quickly. Either of these situations can result in a stall if the situation gets too out of hand.

Many pilots are also concerned about overriding the trim that was set during approach. As you add engine power, the plane will have a tendency to pitch nose up as a

Fig. 6-17 *Effects of indecision during go-around.*

result of the trim setting. You should use forward pressure on the control yoke to overcome this pitch-up tendency; the forward pressure will not damage the trim or flight control system. Once you have your airspeed and rate of climb, retrim the plane to provide a neutral elevator control pressure.

As you transition from a descent to a climb, you will need to use coordinated control inputs, using right rudder to compensate for P-factor to keep the ball centered. You will also need to retract the flaps and landing gear once a safe altitude has been achieved. Retract the flaps before the landing gear as you execute the go-around. The reason for this is because you do not want to retract the landing gear then retract the flaps, which may cause the airplane to settle to the runway if you have insufficient altitude. Flaps also generate more drag than landing gear when they are fully extended, and getting the flaps up can improve your ability to climb. As you climb you should fly to the ride side of the runway to give you a better view of the traffic that may be taking off or landing on it. Once you are back in the pattern, make sure you execute the landing checklist to make sure you do not forget to reset anything to the correct landing configuration.

CONCLUSION

This chapter has covered a great deal of information related to takeoffs and landings. Pattern use, picking your landing point, adjusting your glideslope, and controlling your airspeed and power are all components of a successful landing. Many of the same items are also crucial to a smooth, safe takeoff. As you fly, think about each aspect of a takeoff or landing and critique yourself every time you make a takeoff and landing. If you make it a practice to grade your own efforts, you will find you are able to fine-tune the little items in turning your takeoffs and landings into consistently smooth, safe activities that will impress your passengers.

Several different takeoff and landing scenarios were also covered in this chapter: normal, short field, soft field, and crosswind situations were all discussed in detail. You should practice each of these until you are comfortable with executing them. Begin with an instructor if necessary to help learn the proper use of flight controls, airspeeds, and each aspect of how to successfully complete each takeoff or landing. These can teach you many of the subtleties of how to fly a plane in a controlled manner near the edges of its flight envelope. As you execute each of these activities, keep safety in mind as your highest priority. In many situations you will need to blend these techniques to take off or land. For instance, you may have a soft, short runway you need to take off from, which will require that you use both techniques in concert to get off the ground in the shortest distance possible. Also be aware that conditions may require that you modify techniques to meet a given set of criteria. Each airplane may also be different in how it reacts to a given takeoff or landing. Knowing the airplane and its limitations is very important. What works in one plane may not work for another, so know the planes you fly.

7
Advanced Flight Maneuvers

At this point we are going to move on to a set of three maneuvers that are part of the commercial pilot training curriculum. Each maneuver is designed to further train pilots to fly the airplane near the edges of the flight envelope, dividing their attention between flying the airplane, looking for visual references outside the plane, and checking the flight instruments inside the plane. Very much like ground reference maneuvers, the commercial maneuvers allow the pilot to become more attuned to the plane and the sensitivity of the controls in a changing flight environment and to anticipate control inputs and the result of those inputs.

The first maneuver we will touch on is the chandelle, a direction reversal maneuver that takes the plane from cruise speeds in a climbing turn to just above stall as you exit. As we learn it you will see that the plane is moving through a continuously changing environment and you must be aware of how to adjust your flying techniques to meet the situation. The second maneuver will be the lazy-8. Similar in certain aspects to an S-turn across a road, the maneuver also teaches you to control the airplane through a series of increasing and decreasing airspeeds. Finally, we will look at the steep spiral. Somewhat different than the other two maneuvers, the steep spiral helps you to learn to control the plane's airspeed and bank angle in a flight environment that can seem very unusual to students.

As you read about these maneuvers, think about the control inputs, the changing feel of the pressures on them. You should also think about how you will divide your attention between the instruments and the view outside the airplane. Commercial flight maneuvers will help you to become a more proficient, capable pilot. Learning how to execute each of them in a professional manner will boost your level of confidence and understanding of the airplane. Let's begin with the chandelle.

CHANDELLE

The chandelle was invented during World War I as an escape maneuver by a British pilot by the name of Chandelle. The idea was to reverse course and gain as much altitude as possible during the process to put the pilot in a better position from which to engage the opposing pilots. The FAA terms a chandelle as a maximum fight performance maneuver. During the course of a chandelle the plane should gain the greatest amount of altitude possible for a given degree of bank, and without stalling. While you may learn to anticipate the normal amount of altitude you can gain from a particular airplane, factors that include the bank angle, density altitude, and others will affect the actual amount of altitude gained on a given day. When you are executing a chandelle, you are not graded on the amount of altitude you gain, but on how you fly the maneuver.

Figure 7-1 illustrates a chandelle. To enter the maneuver you want the plane to be in level flight, at cruise power settings, with the landing gear and flaps retracted. Your airspeed should be at normal cruising speeds, or the airplane's maneuvering speed, using the lower of the two. From this straight and level configuration you want to enter a bank angle that is appropriate to the airplane being flown for the maneuver. The lower the performance of the plane you are flying, the shallower the bank will be. In any event, the bank angle should not exceed thirty degrees and should be coordinated from the beginning to the end of the chandelle. After the bank is established, you will want to enter a climbing turn, gradually increasing the pitch angle at a constant rate through proper use of the elevator control. The greatest pitch angle should be achieved as the plane reaches ninety degrees of heading change. As you enter the climb and continue to increase the pitch angle, you should gradually add in engine power. If you are flying an airplane equipped with a fixed pitch propeller, avoid exceeding the red line on the engine r.p.m.'s. Depending on the plane, you may be able to add up to full power during the first ninety degrees of turn. If the plane has a constant speed prop, you may not need to add any power, although it may be beneficial to set the plane up to a climb configuration on engine power and then add additional throttle as the maneuver progresses.

The bank angle should remain constant through the first ninety degrees of a chandelle. For example, if you initially establish a thirty-degree bank as you enter the maneuver, you will want to maintain that bank angle through the first portion of the chandelle. Even though the bank angle should be constant, you will need to adjust your control inputs as the maneuver progresses to hold a fixed bank. As the airspeed decreases and the flight control effectiveness changes, you will need to modify the amount of rudder,

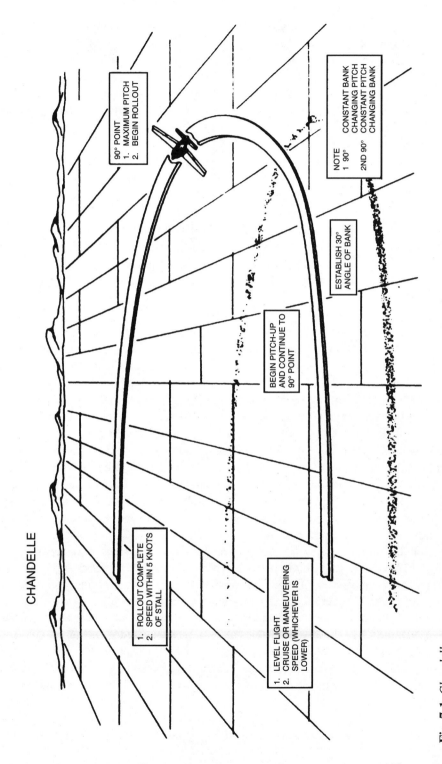

CHANDELLE

90° POINT
1. MAXIMUM PITCH
2. BEGIN ROLLOUT

NOTE
1 90° — CONSTANT BANK
CHANGING PITCH
2ND 90° — CONSTANT PITCH
CHANGING BANK

ESTABLISH 30°
ANGLE OF BANK

BEGIN PITCH-UP
AND CONTINUE TO
90° POINT

ROLLOUT COMPLETE
SPEED WITHIN 5 KNOTS
OF STALL

1. LEVEL FLIGHT
2. CRUISE OR MANEUVERING
SPEED (WHICHEVER IS
LOWER)

Fig. 7-1 *Chandelle.*

191

aileron, and elevator control inputs to maintain the proper bank and pitch angles. After the bank is established, there will a tendency for it to increase and it will be necessary to watch for and compensate with your controls to hold the correct bank.

As you reach the ninety-degree point in the climbing turn, you will want to roll out of it at a constant rate while maintaining a constant pitch angle. This is the opposite of the first ninety degrees of the maneuver, during which we held a constant bank and increased the pitch angle at a constant rate. We are now preparing to exit from the chandelle and setting up the second half of the maneuver. As you roll out of the bank, more lift will be available—that lift vector no longer being used to turn the airplane. Depending on the plane, it may be necessary to reduce the amount of elevator input to prevent the nose of the plane from pitching up as the bank is reduced. You will also need to add right rudder throughout the maneuver to keep the ball centered and compensate for the left-turning tendencies that will become more prevalent. In chandelles to the right, I have found that pilots often have trouble anticipating the amount of right rudder they will need to keep the ball centered. The ball often slips over to the right, and the pilot has to force himself or herself to keep pushing on the right rudder to get the ball back to center.

As we prepare to exit a chandelle we are going to be about five knots above stall speed, the controls becoming less effective as the airspeed slows. To help roll out in a coordinated manner, and to use the controls properly in this portion of the flight regime, you will need to adjust how you fly the airplane as you roll the wings back to level. The aileron lowered into the slipstream as you roll out of the bank creates additional drag and causes the plane to yaw in that direction. You must be prepared for this yawing tendency and anticipate the correct amount of rudder that will be needed to keep the ball centered and the turn coordinated. A rollout to the left can often be done with little left rudder, the effect of aileron drag and torque tending to neutralize each other. In fact, just releasing some of the right rudder you are holding because of the left-turning tendencies is often enough to keep the ball in the middle of the turn and bank indicator (*Flight Training Handbook*, p. 162). When rolling out to the right, it may be necessary to add a great deal of right rudder, since you are already trying to compensate for the left-turning tendencies. In higher-horsepower aircraft, this can really surprise a pilot new to the maneuver.

As you roll the wings to level, you should be just above stall speed, with the airplane's heading 180 degrees from the entry heading. The ball should be centered throughout the chandelle. Once you have established that you have control of the airplane after completing the turn, you will need to transition back to level flight. Depending on the plane you are flying, it may be necessary to add additional power, then slowly lower the nose to let the airspeed build. Once the airspeed begins increasing, slowly reduce power until you are back to a cruise configuration.

Pilots tend to make several common mistakes when practicing chandelles. The first is that they often fail to maintain a constant bank in the first portion of the maneuver, or constant pitch during the second half of it. A very common error is not keeping the ball centered during the entire chandelle, especially in those to the right. Some pilots are also so intent on watching the instruments to maintain the pitch and bank that they fail to look outside the airplane. I feel being able to know what is going on during this maneuver with minimal reference to the instruments is important. The point of the commercial maneuvers is to get pilots to build a greater feel for the plane, and to divide their attention between the cockpit and what is outside. With practice you will be able to know what the bank angle is, and what the pitch angle is, without the instruments. After you work at it you will be surprised at how well you can feel what the airplane is telling you as the chandelle progresses. Now let's move on to lazy-8's.

LAZY 8

The lazy 8 is yet another maneuver designed to teach pilots how to control the airplane through varying airspeeds and attitudes while dividing their attention between flying the aircraft, looking outside the airplane for orientation, and checking the flight instruments. The "8" reference comes from the fact that the longitudinal axis of the plane traces a horizontal figure 8 as it flies through the maneuver. As the maneuver progresses, you will find that due to varying airspeeds the use of flight controls must be adjusted to control the airplane in a positive, exact manner.

Figure 7-2 illustrates the lazy 8, showing how the plane makes two 180-degree turns in opposite directions while executing a climb and then an equal descent in each half of the maneuver. The airplane enters the maneuver from straight and level, but at no time after that is the plane flown in a straight and level attitude. As the plane transitions from the first half of the lazy 8, it rolls immediately into the second half of the maneuver. Before you begin the maneuver, or any other maneuver for that matter, be certain you clear the airspace around you for other traffic. This is especially important with maneuvers that result in climbs or dives, since it may be more difficult to see other aircraft that may be above or below you.

I like to teach students to begin lazy 8's on cardinal headings: north, west, south, or east. This makes it easier for them to judge their progress through the maneuver and to know when they have completed each 90 degrees of turn. You should also pick out reference points that are at 45 and 145 degrees from the initial heading to be used to gauge whether you have established the correct bank angles as the lazy 8 progresses.

As you can see from the diagram, the maneuver is entered from straight and level flight. Like the chandelle, the airspeed should be either cruise or maneuvering airspeed, whichever is lower. You will see how the lazy 8 is similar in many respects to the chandelle as we continue to break the maneuver down. As you begin the maneuver, pitch the nose of the plane up at a constant rate through use of nose-up elevator. At the same time that the nose of the plane pitches up, you should begin a turn in the direction of the 45-degree reference point. The maximum pitch angle should be reached at the 45-degree point, with the bank angle at 15 degrees at the same time. Note that as the airspeed drops off due to the increasing pitch, the rate of turn will increase unless you plan and compensate for this tendency. Use a slow roll rate into the maneuver to counteract this increasing rate of turn. You should also make sure you keep the flight controls coordinated throughout the maneuver. The falling airspeed and high pitch angle will cause the plane to have left-turning tendencies, while the flight controls will become less effective.

LAZY EIGHT

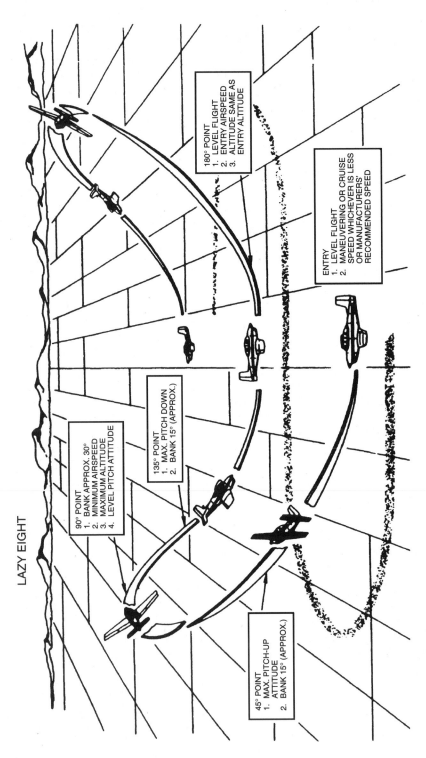

180° POINT
1. LEVEL FLIGHT
2. ENTRY AIRSPEED
3. ALTITUDE SAME AS ENTRY ALTITUDE

ENTRY
1. LEVEL FLIGHT
2. MANEUVERING OR CRUISE SPEED WHICHEVER IS LESS OR MANUFACTURERS' RECOMMENDED SPEED

90° POINT
1. BANK APPROX. 30°
2. MINIMUM AIRSPEED
3. MAXIMUM ALTITUDE
4. LEVEL PITCH ATTITUDE

135° POINT
1. MAX. PITCH DOWN
2. BANK 15° (APPROX.)

45° POINT
1. MAX. PITCH-UP ATTITUDE
2. BANK 15° (APPROX.)

Fig. 7-2 *Lazy 8.*

194

Use sufficient rudder to keep the ball centered, and aileron to maintain a constant roll rate into the turn.

When you reach the 45-degree point of the maneuver, you will have reached the maximum pitch angle but will continue to slowly roll the airplane toward the maximum bank angle of 30 degrees. At this point the nose of the plane should be allowed to slowly drop toward the horizon as you continue to turn toward the 90-degree reference point. Remember that you will still be increasing the bank angle as you transition the pitch angle down. Many pilots first learning lazy 8's tend to freeze the bank angle as they drop the nose toward the horizon. Be prepared for the continued loss of airspeed and the effect that will have on the controls. It may be necessary to continue to increase the amount of right rudder you are using and to roll in a slight amount of opposite aileron to control the angle of bank. At the 90-degree point of the lazy-8, the nose should be level with the horizon and the bank angle should be at 30 degrees. The airspeed should be 5 to 10 knots above stall speed.

At this point you will continue into a descending turn, allowing the nose of the plane to drop below the horizon at the same rate it was raised during the initial pitch-up. The plane's nose will reach the lowest point at the 135-degree point of the turn and should match the angle of pitch-up that the plane achieves at the 45-degree point of the turn. The angle of bank should be gradually reduced as you pass 90 degrees of turn, with the bank angle at 15 degrees at the 135-degree point of the turn. As the airspeed increases during the second 90 degrees of turn, you will need to gradually reduce rudder and aileron inputs to maintain coordinated use of flight controls and to hold the correct angle of bank.

As the plane passes through the 135-degree point of the lazy 8, you will begin to gradually raise the nose back toward the horizon and continue to roll the wings level. The plane's nose and wings should just reach level with the horizon as the plane hits 180 degrees of turn. The airspeed at this point should equal the entry airspeed. The airplane is then immediately flown into a turn in the opposite direction, using the same pitch and bank angles at the 45-, 90-, 135- and 180-degree points of the turn.

Because of the changing bank angles, pitch angles, and airspeeds, you will find that the use of controls changes throughout a lazy 8. Like the chandelle, the amount of right rudder needed during a turn to the right will be greater than that needed during a turn to the left. Torque will also have a stronger tendency to increase the angle of bank when you are turning to the left, and to reduce the bank angle when you are making turns to the right.

As you become more proficient in executing lazy 8's, you can increase the angle of bank to more than 30 degrees to make the maneuver more challenging. Be aware that as the bank angle becomes steeper, the use of controls will become more important. The airspeed will have a tendency to decrease and increase at a higher rate, and it will be easier to let the airplane's rate of turn get away from you. However, I get a real kick out of doing lazy 8's in the Pitts Special I fly, letting the wings roll to 90 degrees at the 90-degree point of the turn. Never take the maneuver to this extreme unless you are flying an aerobatic airplane and have parachutes on though. A lazy 8 can get away from you if you are not familiar with how the plane will react, and if you take the bank angle too high, the nose can drop rapidly, letting the airspeed build to an excessive amount.

Pilots frequently do not use the correct pitch angles, either pulling the nose up too quickly or not fast enough during the first 45 degrees of turn. If the nose gets too high, the airspeed drops off quickly and is too slow at the 90-degree point. The rate of turn is also affected and can make the maneuver less symmetrical. When the pitch angle is too low, the airspeed remains higher than it should at the 90-degree point, and the rate of turn is once again affected. Pilots seem to have fewer problems with the bank angle, but it may take practice to get the bank angle to be 15 degrees at 45 degrees of turn, and 30 degrees at the 90-degree point. Getting the pitch above and below the horizon to match also takes practice, as does ending up with the exit airspeed being the same as the entry airspeed. Be aware that in the right-climbing turn portion of the maneuver you will have slightly crossed controls to prevent the airplane from overbanking. This will require that you use some left aileron pressure to prevent overbanking to the right, and right rudder to compensate for the effects of torque.

I really like flying lazy-8's. On a nice day you can do a whole bunch of them in a row, letting a rhythm build up as the nose rises and falls and the bank increases then decreases. When you become proficient at lazy 8's, you can sense the plane moving around you, the change in how the flight controls feel, and the decreasing and increasing airspeeds. This is really a fun maneuver to learn and really helps you build a feel for the airplane and how to control it in varying flight conditions. Even if you never go for your commercial rating, I highly recommend getting some training in executing it just for fun. Now let's take a look at steep spirals.

STEEP SPIRALS

Steep spirals combine several aspects of maneuver that we have covered so far in the book. In some respects it is like a turn about a point. You want to maintain a constant radius from the point you use as the center of the maneuver. To perform the maneuver correctly you must also maintain a constant airspeed. The airspeed control for steep spirals will be accomplished with the same airspeed control techniques we learned for takeoffs and landings. Pitch controls airspeed. Power controls rate of descent. Steep spirals also force you to divide your attention between flying the airplane, watching the reference point, and monitoring the instruments.

The steep spiral is illustrated in Figure 7-3 and is an excellent maneuver to teach pilots how to make descending turns around a fixed point. This skill can be handy not only in the pattern, or when descending toward and airport, but especially in emergency landing situations. Using steep spirals can help you set up to enter downwind, base, or final to the field you may need to land in during a forced landing situation. As you look at Figure 7-3 you can see that the plane maintains a constant radius around the tower as it descends. Through proper use of a steep spiral, you can use your altitude to your advantage if you are ever faced with a real engine-out situation.

The steep spiral is entered heading into the wind at an altitude that will allow three complete turns before you exit. Since you do not want to descend below safe altitudes, this means you must enter the maneuver at an altitude that will let you safely complete the steep spiral. Entering the maneuver at too high an altitude can also be a problem,

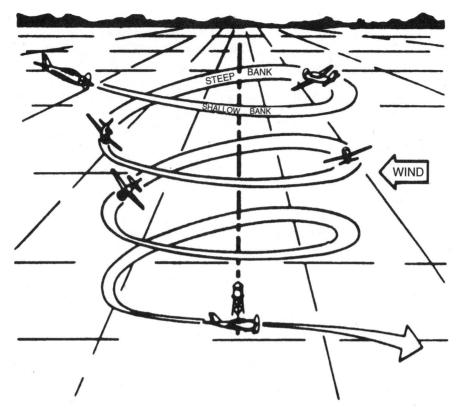

Fig. 7-3 *Steep spiral.*

making it more difficult to see the reference point you are using as the center of the turn. The radius you use during the turn should be such that you will need a 50 to 55-degree bank at the steepest point in the turn. As with ground reference maneuvers, and in particular turns about a point, the bank will be steepest when the ground speed is highest and shallower as the ground speed slows. If you look at the figure, you can see that when the plane is on a downwind heading, the bank is at its steepest angle and shallowest when the plane is heading into the wind. Moderate bank angles will be used when the airplane is perpendicular to the direction of the wind.

Power should be at idle during the maneuver, allowing for a steep rate of descent. Because prolonged idle can cause the engine to cool to the point that it will no longer run, you should "clear the engine" once or twice during the steep spiral by advancing the throttle to cruise power and letting the engine warm under its own power. This will change the glideslope slightly, but turning a practiced force landing into a real engine-out situation can make for an exciting lesson. As you advance engine power, you will need to adjust your pitch angle to maintain a constant airspeed. To help reduce the impact of increasing the power, you might try increasing power as you head into the wind. I recently attended a flight instructor renewal class. During one session the instructor showed a video of a student and his flight instructor practicing power-off stalls. The student had brought along a video camera and placed it behind and between the pilot and copilot seats to record the lesson. Carburetor ice formed during one of the power-off stalls, and when the student recovered from the stall the engine stopped. The student and instructor did an excellent job of making a forced landing, ending the flight with no injury or damage to the aircraft. But this shows how important it is to clear the engine during any power-off maneuver you may be practicing.

Before beginning a steep spiral, find a good reference point, such as a road intersection or water tower in an uncongested area, then clear the airspace of other traffic. Reduce your power to idle and establish your airspeed. V_x is a good airspeed to work with during the maneuver. However, if the aircraft manufacturer recommends a different value for the best glide speed, use that instead. As you come abeam the reference point, enter the turn using a shallow bank to begin with and noting your distance from it. As the turn progresses you will gradually increase the bank until you are at approximately 50 to 55 degrees on the downwind portion of the turn. At that point you will gradually reduce the bank angle until it reaches the shallowest bank on the upwind heading.

As the bank angle changes you will also need to adjust the pitch angle to hold a constant airspeed. During a steep spiral the airspeed will have a tendency to increase and decrease with changes in the bank angle if the pitch angle is not adjusted accordingly. When the bank angle is steeper, the pitch angle must be reduced. As you shallow the bank angle you will need to increase the pitch angle to maintain the correct airspeed.

Because of the relatively slow airspeed you will need to be aware of control effectiveness and maintain coordinate use of the controls. It is not uncommon for pilots to cross-control to some degree as the bank becomes steeper. Instead of using the elevator, ailerons, and rudder in a coordinated manner, pilots learning the maneuver may hold "top" rudder to help reduce the backpressure necessary to hold the correct pitch angle. As the maneuver progresses you will continue the changes in bank and pitch angles to hold a constant radius and airspeed. You should roll out after completing a minimum of three turns, while still at a safe altitude. The rollout point should be a predetermined heading or reference point you picked before entering the steep spiral.

Many students have trouble maintaining a constant radius during the maneuver. Because of the steeper banks, it is easier to let the radius change rapidly if you do not anticipate the effects of the wind, and adjust your bank angle accordingly. You will probably find that this is more challenging than turns about a point from that respect. Airspeed control will also be a greater challenge as a result of the bank angles. It is not unusual for pilots to let the airspeed build as the bank angle increases, or let it get too slow as the bank angle is reduced.

The steep spiral puts to use many of the skills we have discussed so far in the book. It tests your ability to judge the effects of wind, how to compensate for the wind with varying bank angles, and how to hold a constant airspeed. It also divides your attention between watching the reference point, the instruments, and knowing what your relation to the wind direction is. As you master this maneuver, you will find that it helps you to further refine the aspects of flying that will help you to become a safer, more proficient pilot.

STEEP POWER TURNS

The last maneuver we will cover in this chapter is steep power turns. Over the years I have practiced this maneuver with quite a few pilots; it's one of the first I have aerobatic students do to learn the feel of the plane. It also gives me a way to judge a pilot's ability to use coordinated controls in high-performance maneuvers. The majority of pilots I have flown with do not execute steep power turns very often, if ever, and keeping the controls coordinated can sometimes be a problem. We will discuss the common problems pilots have with steep power turns at the end of this section, but first let's look at what the maneuver actually is.

Figure 7-4 illustrates a steep power turn, and it's a very straightforward-looking maneuver. Here are the criteria that determine if the turn is "steep." The bank angle must be large enough that there is a tendency for "overbanking," or the bank actually steepens on its own and you must use opposite aileron to prevent the bank from becoming greater. There is also a relatively high load factor imposed on the wings as a result of the g-forces the turn imposes. Like the other maneuvers we have discussed in this chapter, the purpose of a steep power turn is to teach you to use smooth, coordinated controls inputs and to be able to divide your attention between flying the airplane and looking outside for reference and inside at the instrument panel.

Looking at the figure, you can see that the plane essentially enters a turn. Unlike ground reference maneuvers, you are not using a reference point at the center of the turn and you do not need to be at pattern altitudes. In fact, until you become proficient at steep power turns you will want to maintain a higher altitude to avoid problems if you should stall the plane accidentally. For most general-aviation aircraft, the bank angle will be somewhere around 50 to 60 degrees. You do not want to exceed the recommended bank angle due to the load factors it may impose on the plane's structures. As

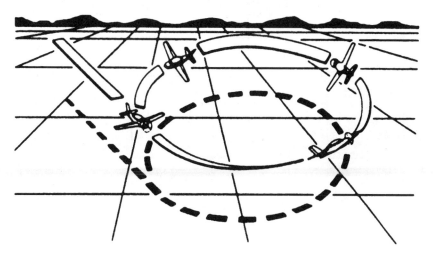

Fig. 7-4 *Steep power turn.*

the bank angle increases, the g-loads on the airplane increase. Figure 7-5 is a table that shows the increasing g-loads generated at a givenbank angle. Once the plane exceeds 60 degrees of bank in level flight, the g-loads increase very rapidly, with the possibility of exceeding some plane's structural capability. At 60 degrees of bank, the g-load is 2 g's. At 70 degrees it increases rapidly to about 3 g's. This is very near the limitation of most general-aviation aircraft in the normal category, so you must watch your bank angle as you practice steep power turns.

As you enter the turn you will want to be at or below maneuvering speed, V_a, to avoid overloading the airplane's structure. You will likely need to carry cruise power at entry, and you may need additional power as you increase the bank and loads on the plane. Roll smoothly into a 50- to 60-degree bank using coordinated rudder and aileron inputs. A noticeable amount of elevator backpressure will be needed to maintain a level altitude as the bank becomes steeper. If you will recall our discussion of turns, a portion of the lift that was being used to hold the plane up is now being used to turn the plane. The steeper the bank, the greater the lift vector used to turn it. To

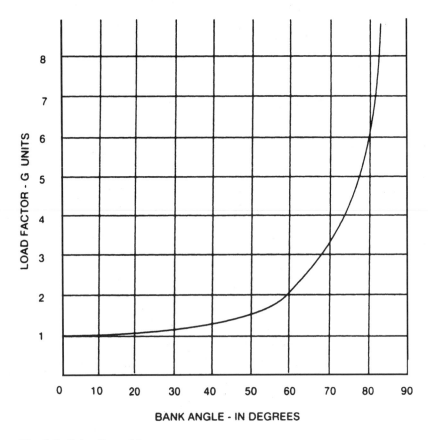

Fig. 7-5 *G-loading table.*

maintain a constant altitude, you will need to increase the overall lift, which is why you must increase the angle of attack and elevator backpressure.

This is where many pilots become uncomfortable with executing steep turns. As the bank becomes steeper and the elevator backpressure increases, so do the g-loads, as compared to normal turns. The elevator backpressure can become quite heavy, and many pilots seem to hesitate pulling that hard. Many find that by using rudder opposite the direction of the turn, they can hold the nose of the plane up and not need so much elevator backpressure. The problem is, the turn is now very uncoordinated, we are at a steep bank angle where the stall speed also increases, and we are in the perfect setup for a high-speed stall/spin situation. Use coordinated control inputs through as you enter the maneuver, and this potential problem can be reduced.

Once you are established in the turn, you will need to use slightly opposite aileron to compensate for the overbanking tendency steep turns cause. Again, as you adjust the aileron, also adjust your rudder inputs. If the plane loses altitude and you are pulling hard on the elevator control, add more power. Just as with our climbing and descending maneuvers, power is used to control altitude. The maximum turning performance for a given aircraft will be reached when the radius of the turn is smallest and the rate of turn is highest. This will vary with the bank angle and airspeed of the plane. You will not be able to tell exactly when the maximum turning performance has been reached, but you can learn to recognize when the bank angle is correct and you are flying in a coordinated manner while maintaining a constant altitude. If power settings cannot be increased and the plane is descending, decrease the bank angle to hold altitude.

As the turn progresses, monitor your heading closely. Many pilots have a tendency to fly past the desired rollout heading because the plane is turning so quickly. As you roll the plane back to level flight, remember to release elevator backpressure and, if necessary, reduce engine power. It is not uncommon for pilots to gain several hundred feet as they roll out of a steep power turn due to the now excess lift that is being produced with the backpressure and engine power that was necessary to hold a constant altitude during the turn. The amount of lead necessary to roll out on heading will depend on the plane and the rate of turn, but be prepared to give a generous lead. Like altitude, until pilots learn to anticipate the rollout point, they often fly past the correct heading for the turn's exit.

Steep power turns also teach pilots to become more comfortable with the plane in unusual flight attitudes. Most pilots rarely exceed 30 degrees of bank, and to nearly double the bank angle while maintaining a constant altitude places them in a new flight regime. However, once pilots master steep power turns, they have a better understanding for how an airplane actually behaves, and they do not have to "cheat" by using uncoordinated control inputs. Pilots also have a tendency to stare at the attitude indicator and altimeter during steep power turns. While these can be used for reference, you should be using the horizon outside the airplane to monitor your pitch and bank angles. You should also look around the sky and horizon, not focusing on just a point ahead of the airplane. As you become more comfortable with steep power turns, you will find that using coordinated flight control inputs during normal flying becomes even easier.

CONCLUSION

This has been a fun chapter because we got into some interesting maneuvers that most pilots don't practice very often. These maneuvers all help us become more aware of the plane and give us the chance to fly the aircraft through varying flight regimes. Learning to recognize by feel the changes that take place in the airplane as the airspeed increases or decreases is invaluable as you fly through the pattern. To know what the plane is doing merely by the feel of the controls can alert you to problems if you become distracted by something inside or outside the airplane. Knowing what a plane behaves like and being able to anticipate how to control it at opposite ends of the flight envelope also make you a smarter pilot. Knowledge is the key to almost anything in life, and flying is no different.

Find a qualified flight instructor you feel comfortable with and go out and learn these maneuvers. While many of them are considered commercial flight test maneuvers, you do not need to get your commercial rating to take advantage of the increased flying skills you can obtain by practicing them. In fact, I'd like to take the recommendation one step further and suggest that you get some basic aerobatic training. Not every pilot needs to learn how to fly aerobatics like Patti Wagstaff or Sean D. Tucker, but learning how to fly a roll or loop can give you a tremendous amount of confidence in unusual flight attitudes. But beyond that, aerobatics are a tremendous amount of fun, and that is what flying should be about for all of us.

8
Emergency Procedures

Most of us are never faced with a true emergency situation in flight; this speaks well of the safety standards incorporated into the design, manufacture, and maintenance of the aircraft we fly. It also is a positive reflection of the flight training the majority of us receive as we progress through our flying careers. But every once in a while something comes along that makes us sit up straight in the seat—a real emergency. Whether it's an engine failure, weather related, or being just plain lost, how we react to these unplanned adrenaline rushes can make the difference between life and death.

In this chapter we are going to look at several different emergency situations. We cannot cover every emergency situation, but we will look at how you can react to emergency scenarios in general to help improve your chances of getting things back under control. As you read this chapter, keep in mind that you are the key to resolving any problem. If you keep your head, keep flying the airplane, and work through the problem, you will improve your chances of a successful end to your flight.

The scenarios we will cover in this chapter will include being lost, getting into unexpected weather conditions, and engine-related emergencies. We will finish the chapter with a discussion of how to avoid accidents and emergencies. It is much easier to stay out of emergencies than to get out of them once you get into a bad situation. The information in this chapter is only a guideline; it cannot be a "cookbook" recipe to everything that flying might throw at you. As we will discuss later in the chapter, poor judgment is what often puts us "in harms way." Through planning and sound judgment we can avoid most situations before they happen.

LOST

I'm willing to bet that nearly all student pilots have at one time during their solo cross-counties thought they were lost or off course. I remember taking a solo cross-country, flying along as the landscape rolled along beneath me. It was a clear blue day and I was feeling very good about being a pilot. All of a sudden I came to a town that wasn't on my chart and that uneasy feeling began to build. I descended to be able to get a look at the name of the town on the water tower but still could not find where I was on the map. I was at the edge of the first chart and began looking on the next chart, all the while circling the town. Finally, I noticed that the real edge of the first chart had been flipped under, and in my haste while looking at the chart I had failed to notice it. I unfolded the chart, and yes, there the town was, right on my course.

You may be asking yourself why I would relate such an embarrassing story to you. Because these are just the types of things that cause pilots to get into trouble while they are flying cross-country. If I had panicked and struck off in a direction based on a hunch instead of trying to understand what was going on before I made a decision on the next step I should take, I could have ended up really lost. Instead I was exactly on course but had misused the charts. My mistake, not the charts'.

We can end up lost for many reasons: stronger winds than expected can blow us off course; radio navigation equipment can fail or give erroneous readings; we could have picked the wrong heading to fly through misreading the map as we planned our flight. Using an out-of-date sectional chart could have us dialed into a VOR using the wrong frequency; there are more reasons for getting lost than can be listed. Whatever the reason, though, we need to cross-check what we are doing often enough to assure that we do not have a single point of failure as we plan, then execute, our cross-country flights.

Use current charts and weather information as you plan your cross-country flights to help assure that your planning is accurate. As you fly, frequently verify that the radio navigation equipment is working properly by cross-checking the information against what you see outside the window and on your sectional chart. By using several methods of figuring out your position, you can prevent having one failure cause a real emergency. On VFR days, I fly using pretty much only a compass and the sectional chart. As long as I can see the ground and I don't lose the chart, I know exactly where I am. But even with that I will cross-check my position with a VOR every once in a while to verify that I am really where I think I am.

But what happens if you have done all of these things and you suddenly do end up lost? What can you do to help get yourself "found" again? First and foremost, keep flying the airplane. If you quit flying the airplane and start to let the heading or altitude get away from you, you have already lost the battle. If you are unsure of your position you actually have several options open to you. Many of us do not want to call and ask for help, egos being what they are, but that is a very easy way of getting out of the situation. There are very few areas in the country any more that are not in controlled airspace. If you are equipped with a radio and a transponder, you can ask them to help locate you, and the controllers will be more than happy to help you. If you are equipped with a VOR you can cross-check your bearing from two or more VOR stations and triangulate your position. (Make sure your map is open all the way and not folded under!) More and more aircraft and pilots are now flying with GPS, which is an incredible piece of equipment. They often have a "nearest" function, which gives you headings and distance to the nearest airport. If you have a properly functioning GPS on board, you should not get lost, but if you do it should be able to help you get to an airport with little difficulty.

The worst lost situations are when you are low on fuel, with poor weather conditions at night. Proper management of fuel and monitoring weather as your flight progresses can help you stay out of this particularly bad scenario. You will see in the "pilot judgment chain" later in this chapter how we often put ourselves into situations like this by falling prey to factors we can actually control. "I've got to get there" will often make us do things we know we should not. Who among us has not thought, "I can get there before things get too bad," trying to squeak into an airport before nightfall or weather makes the situation dangerous. "I want to sleep in my own bed tonight," has probably caused more cross-country related accidents than we realize. Don't set yourself up for low fuel, poor visibility, cross-country flights trying to stretch the range of the plane. I do not let the fuel get below half tanks when I fly because I don't want to worry about how I'm going to make it to the next airport. Common sense as we fly cross-country can help keep us from ending up lost, with low fuel and few options for getting things worked out.

The best thing you can do on any flight is to monitor your position closely using several methods. This can help you avoid getting into trouble. If you do lose your position, don't panic. Calm, clear thinking is your best ally and can get you out of the scrape. As soon as you begin to suspect that you may be out of position, make a plan for working out where you are, and stick to that plan. Think before you act, and you've won half the battle.

WEATHER-RELATED EMERGENCIES

Weather-related emergencies can be very dangerous situations that you need to be aware of and prepared for. A weather-related emergency can be any situation involving weather that can range from flying into unexpected IFR (instrument flight rules) when you are only VFR rated, to thunderstorms that you encounter, to extreme crosswind conditions that may exceed the plane's or your capabilities to handle. Heavy icing, hail, and microbursts (also known as wind shear) are yet other examples of the types of situations that Mother Nature can throw at you while you fly.

We cannot cover all weather emergencies, but we will discuss how to avoid them and some general guidelines that may prove useful if you do get yourself into a situation. You will notice a recurring theme in this chapter, and that is to fly the airplane and think clearly. If you can do those two things, you will improve your chances of getting out of trouble. The last thing you ever want to let happen in an emergency situation is to lose control of yourself or the airplane.

The best way to deal with weather-related situations is to avoid them. Getting timely weather briefings from Flight Service can give you an accurate picture of weather that you can expect to encounter when you fly. After you take off, you should also check with Flight Service via the radio en route to obtain weather updates along your route of flight. These weather briefings and updates can provide you with information that is valuable to your decision-making process. Now comes the hard part for some pilots—changing their course, or landing, to avoid questionable weather situations. When we fly cross-country it is normally because we want or need to get to our point of destination for a reason. Business, family, and other reasons motivate us to make these trips. For these very same reasons, we may hesitate to delay our arrival time when faced with a change in weather, especially after we are already in the air and on our way.

It is at these times we might put ourselves into a situation that may be difficult to deal with safely. The "I've got to get there" syndrome kicks in, and we forge on, hoping the weather will not be as bad as predicted. I made similar decisions early in my flying career because I wanted to get where I was going. I was flying from Minnesota to Illinois during the winter. I checked with Flight Service before takeoff and the weather was forecast to be VFR all the way along the route of flight. The flight was about three hours long, and I made the mistake of not bothering to get an update while en route. As the flight neared central Wisconsin, the ceiling started to drop, and it began to snow. I kept descending to stay under the clouds, but altitude and visibility were rapidly diminishing. As I flew by Madison, Wisconsin, I was getting very uncomfortable with the now very low visibility and heavy snow, and I obtained a special VFR into the Madison airport. This situation left a lasting impression on me. If I had checked with Fight Service as I had flown along, I would have known about the deteriorating weather, instead of getting caught in a situation I would rather have avoided. An Aztec that landed shortly after I had taxied up at Madison had so much ice on it that the pilot had to maintain an extra 20 knots of airspeed to keep from stalling during the approach. I was lucky the weather did not turn to IFR and put me in a very bad situation, but I learned a lesson from the flight: check the weather, and check it often.

What should you do if you encounter severe weather? Well, there is no one right answer, but there are several things you should do to give you the information you need to improve your chances of making it to a safe landing. First, fly the airplane. Second, think about what is going on. Are you in thunderstorms, heavy icing conditions, or unexpected IFR conditions? In other words, what is the real problem? With weather you can sometimes turn around to get out of it, and that may be the best thing to do in certain situations. Talking to air traffic control may be a good place to start as well. In many cases, once you establish contact with them they can vector you around or through the weather. You may think the thunderstorm cell you are in is very small, but air traffic control may be able to give you the safest route to fly to assure your safety. So call them as soon as things start to deteriorate. Once again, ego gets in the way of many pilots making that call, but it can be the safest thing to do.

There are some standard steps you can take as you fly in weather that can help improve the safety of the flight. If you are in icing conditions, talk to air traffic and let them know what is going on, then ask for an altitude that will provide warmer temperatures. If you are in unexpected thunderstorms, let ATC know that. It may be impossible for you to hold altitude with updrafts and downdrafts that can be present around a thunderstorm cell. In these cases, fly the airplane at or below maneuvering speed, and maintain attitude. There is not much any plane can do to overcome the up and downdrafts, but if you keep the plane level and don't let the airspeed get away from you, it improves the chances that you will fly through the cell safely. It can be very rough inside a thunderstorm, and you want to make sure your seat belt it very tight around you. Even then you may find your head banging up against the cabin ceiling.

Being caught as a VFR pilot in IFR conditions can be very unnerving. As a VFR pilot you use the instruments, but you are very reliant on the ground for navigation and location fixes. Losing that reference forces you to fly strictly on instruments. If you find yourself in this situation, keep flying the airplane, maintain a level attitude, and contact ATC, then let them know what your situation is. They would much rather help you get to an airport safely than have you poking through clouds and not be communicating with them. As you fly, do not fall into the trap of focusing on just one instrument. Pilots will often focus on just the artificial horizon, viewing it as their only way to hold the airplane's attitude. Scan all of the instruments, using that information to fly the airplane on the correct heading and altitude. Maintain a safe airspeed, altitude, and heading, and you've got the plane under control.

Weather is always a major factor when we fly. If you respect that and use the resources of the FAA to determine what the weather is predicted to do when you fly, you will help avoid any weather-related surprises. But Flight Service is not able to forecast weather with 100 percent accuracy. Conditions can change rapidly, and you should combine what you have been told with what you see. If you begin seeing towering cumulus clouds on a warm, summer afternoon, it may be a precursor to thunderstorms. If the temperature is dropping as you encounter precipitation, you may end up in icing conditions. The combinations are endless. Notice what is going on and recognize that weather can change rapidly and that you may need to adjust your flight plans accordingly. Getting to your destination a few hours later because you waited for weather, or diverted around it, is much better than the alternative of not making it at all.

ENGINE PROBLEMS

Engine problems can come in a variety of manners: you can have partial power loss, oil pressure loss, or fuel pressure loss. You can lose the engine altogether, or if you are flying a constant-speed prop, you can lose the ability to change the propeller setting. Engine problems can occur during takeoff, at cruise altitude, or during landing. They can also occur at night or during the day. How you deal with an engine problem in the first few seconds of the emergency will play a great role in the outcome of the situation. In this section we will look at a number of engine-related scenarios and some guidelines for dealing with each of these situations.

It's been previously stated in this chapter, but these are only suggestions for how to deal with various engine problems. No book can give you a complete set of emergency procedures or anticipate every contingency you may have. Think about the various situations as you read the following section, and try to develop further guidelines. You may find that it could be beneficial to create an emergency checklist that you could store in the plane and pull out when a real engine emergency takes place. This could be used to supplement the checklists in the pilot's operations manual for the plane you fly. Let's begin with an engine loss at takeoff, one of the more interesting engine-out situations a pilot can encounter.

Engine Loss during Takeoff

Losing your engine at low altitude and at a relatively low airspeed is a circumstance that demands that you quickly recognize the situation, set the airplane up for its best glide airspeed, and pick out a place to land. Depending on the altitude you lose the engine at, this could all take place very quickly, with the only spot that you can land being straight ahead of the airplane. If you have more altitude, you may be able to scan to the left and right for the most suitable landing spot available, while with enough altitude you may be able to return to the airport for a landing. The final scenario is suggested only when the airplane is at least 500 feet above the ground and depends on the airplane you are flying, the weather conditions, and a host of other factors.

In this section we are going to review a number of factors associated with engine loss and how to make a landing from an engine-out situation. We often receive warning signs that the engine is having problems. If we are aware of those signs and react to them, we can often avoid having a real emergency situation develop. We will discuss a number of those signs, what they mean, and how we might deal with them. If an actual engine-out situation occurs, you need to know the limitations of the plane you are flying and how to quickly pick out a suitable landing area. Pilots faced with an engine out will often try and pick a landing area that is too far away or that requires too great a turn to reach it. We will also discuss how to limit your landing sites to those that are most suitable and provide the greatest chances for success. Let's begin with warning signs of a potential engine problem.

Engine-Out Warning Signs

Very few engines fail with a sudden, catastrophic breakdown that provides absolutely no warning that a potential problem exists. A crankshaft, piston, or other major engine component can fail suddenly, but these cases are very few and far between. And few of these types of failures are actually due to a defect in the component. Improper engine operation on the part of the pilot can cause detonation, high operating temperatures, and loss of oil pressure that often contribute to these major engine component failures. Knowing proper engine operating procedures for the plane you fly can reduce the chances that a pilot-induced component failure might take place.

You should review the operations manual for the plane you fly to learn the proper procedures for setting engine r.p.m.'s and manifold pressures and leaning the mixture. Excessive leaning can cause an uncontrolled, explosive burning within the cylinder known as *detonation*. Figure 8-1 shows that during normal engine combustion, the forces within each cylinder generate powers that push the piston downward, resulting in the crankshaft and propeller turning. But when the mixture has been leaned excessively,

NORMAL COMBUSTION

NORMAL BURNING

EXPLOSION

DETONATION

Fig. 8-1 *Normal versus abnormal engine combustion.*

the fuel/air mixture in the cylinder burns too rapidly and in the wrong burn pattern. This can result in explosively high pressures building within the cylinder that can eventually damage the piston, valves, and cylinder assembly. Detonation can cause failure of these components that can result in either partial power loss in early phases or complete engine failure in severe cases.

Your airplane's operations manual will provide guidelines for power settings and leaning the mixture for various altitudes and configurations. Learn those power settings and use them correctly to reduce wear and improve the lifetime on the engine. Some pilots make the mistake of using incorrect manifold pressures for a given r.p.m. setting, which can contribute to excessive wear and possible engine failure. Others fail to correctly manage the mixture setting, either consistently leaning the engine too much for a given power setting, or not adjusting the mixture as power or altitude changes take place. Engine failures aside, your engine is one of the most expensive components on your aircraft. Through proper utilization in accordance with the manufacturer's recommendations, you can extend the engine's life and reduce your operating costs.

Now let's take a look at some of the warning signs that you may be having engine trouble, and what you might do to help improve the situation. Again, these are only suggestions;

the engine and other conditions could very likely warrant other actions. First, the runup is one of the best things you can do before takeoff to make sure everything is running correctly. By checking the magnetos, vacuum and oil pressures, oil, and cylinder head temperatures, you can determine if there are any potential problems before you get in the air. Many pilots just sort of breeze through the runup before they take off, not really noticing what the readings on engine instruments are or knowing what the correct ranges are. I'm a firm believer in really paying attention to the checklist and verifying proper engine operation.

About three years ago I did a runup in the Pitts Special I fly, getting ready for an aerobatic flight. I taxied out to the runway, started the runup, and did the magneto check. Much to my surprise, one of the mags was dead, and I had to taxi back to the ramp to see what the problem was. It turned out to be a broken P-lead on the mag, and in about an hour I was ready to fly. But if I had not done a proper runup, I could very well have ended up in a serious situation.

What other things might show up during the runup that could be indications of potential engine problems? Oil pressure is a very good indicator of engine health. If your oil pressure is low, or high, you may have problems with the oil pump, oil screen or filter, or oil levels. I talked to a pilot who had had his oil changed and the screen checked. The oil screen housing had been improperly reinstalled, and oil began leaking from the engine. During the runup, the pilot noticed the low oil pressure because of the leak, and taxied back to the hanger, only to discover what the cause of the problem was. If he had not noticed this problem before takeoff, complete engine failure was almost certainly guaranteed. If your plane is equipped with a constant speed prop and you find that the propeller cycles slowly, or not at all, or that oil pressure fluctuates beyond norms, you may have other oil-related problems. A leak in the propeller hub or governor may result in oil pressure problems as well.

A rough engine or excessive r.p.m. drop during the magneto check may be a signal to mag-related problems. This may also be an indication that you might have fouled spark plugs, excessive oil passing by the piston rings, or a bad spark plug lead. Carburetor ice may also cause a rough engine during runup, so you may want to verify that carb heat was applied during the runup. You can also try leaning the mixture somewhat to make sure you did not foul the plugs during start-up or taxi, but if this does not work you should have a mechanic look the engine over to determine what the problem is. Low or high cylinder head temperatures may be an indication of improper combustion within a cylinder. If you find that this is the case, check the mixture to be sure the setting is correct. You might also cross-check this information with the mag check to see if you have a rough-running engine that the cylinder head temperature might give additional information about. Fuel pressure should also be checked and verified at the correct pressure ranges. Low fuel pressure could have significant effects on engine performance, especially during full throttle operations.

In-flight you might notice similar changes taking place for all of these engine indicators. If your oil pressure drops, or the oil temperature increases, you may have signs of an engine that could be suffering from a variety of problems. Once you notice either of these problems, you should look at other engine-related gauges. Is the cylinder head temperature (CHT) also high? Have you set engine power settings correctly? What are the cowl flaps set at? These are but a few cross-checks you can look at to see what the extent of your engine problems might be. If the CHT is also high, there could be significant engine problems causing the engine to be overheating due to the engine having to work extremely hard. Engine parts failing could be the cause of these indicators. However, the same engine temps could be present if you have left the cowl flap closed during climb or when you have been running the engine at high power settings. If you are flying and see a change in engine operating characteristics, don't focus on just the gauge that is giving you the problematic reading. Look around and see what else is going on with the engine. You may have caused some engine readings through improper settings or configurations, and recognizing that will allow you to fix the problem.

If there are no logical explanations for low oil pressure, or high temperatures, you should head for the nearest airport to land and have the engine checked out. While it may be an inconvenience to make this side trip if the problem turns out to be minor, I'd much rather make a decision on the conservative side than try to find a field to land in if things begin to get worse. No manual or emergency procedure can cover all engine failure situations, but using sound judgment can help you deal with a potential problem and keep it from getting out of control. Now let's take a look at an engine-out emergency landing and the proper way to approach it.

Engine Failures

If you actually lose an engine, there are several steps you will need to take to improve your chances of finding a safe landing area and making it to that landing zone. In this section we will review several concepts regarding forced landings. We will also take a special look at losing an engine shortly after takeoff, a very special engine-out situation. In any engine loss condition, the most important thing you can do is to keep flying the airplane and to keep thinking. Let's take a look at some of the things you might need to think about.

The majority of the flying I do is in the Midwest, which is made up mostly of farm fields and small towns scattered around the countryside. In this area of the country, the average pilot flying a Cessna 172 or similar plane has an almost limitless number of potential forced landing areas to pick from. In other areas of the country, or around large cities, you will have fewer options for emergency landing sites. But as you fly you should always consider what you would do if you did lose an engine. If you start thinking this way, you will be surprised at how it becomes almost second nature to scan the ground below you for landing sites as you fly along. At first it takes some conscious thought, but with a little practice you quickly include this as part of your normal scan while you fly.

An interesting result of doing this for me has been that it strikes me very quickly as I fly over large cities or rough terrain that the number of landing sites becomes very limited. Take downtown Chicago, for instance. Unless you are willing to risk landing on one of the many roadways criss-crossing the city, there are very few choices for an engine-out landing site. Fortunately, the reliability of engines is very high, but this does not mean we do not need to consider what would happen in these situations. In some cases, flying at a higher altitude will provide you with more landing site options, while in other cases you may need to actually fly around a very inhospitable area to avoid having no emergency landing sites. This second option may seem extreme to some pilots, but as a pilot who has had to make a forced landing, I am a firm believer in keeping as many options open as possible.

As soon as you discover you have an engine loss, there are several steps you will want to perform immediately. The first is to set your airspeed for the best glide value. If you refer to the emergency section of the pilot's operations manual or in some cases the airplane's checklist, the correct airspeed will be documented. The reason for establishing best glide is to give you the maximum distance to choose potential landing sites from for a given altitude. If you use an airspeed above or below the best glide airspeed, you will lose altitude more quickly and reduce the gliding distance the plane is capable of. Trim the airplane for the best glide airspeed to reduce your workload and make it easier to hold the correct airspeed.

After the best glide airspeed has been established, look for a suitable landing area. You will have to base this decision on the conditions at the time, such as the terrain you are flying over, your altitude, landing area options within the airplane's gliding distance, and how much area your plane will need to land safely. In a true emergency-landing situation, you may have limited choices, but you want to pick the best landing site available from those choices. You will want an open area that is long enough and offers minimal obstructions such as trees or power lines. If you are landing in a field, you will want to land with the rows, as opposed to across them, so the direction of the field in relation to the direction of the wind will also be a factor. Landing into the wind will be the best choice to help reduce the landing distance as much as possible. Figure 8-2 shows an example picking a field with the furrows running as close to into the wind as possible. If the fields have crops growing in them, the type of crop can also affect your decision for landing. Given a choice between a cornfield and a bean field with the same furrow directions in relation to the wind, you would normally pick the bean field to land in. The corn stalks will be taller, making it more difficult to judge your flare. The corn can also hide obstacles that might cause damage to the plane after touchdown. Additionally, the corn stalks might cause damage to the plane's leading edges as you move through them at a high speed. This can be especially true for fabric-covered aircraft. Picking the right field can help minimize the damage to not only your plane, but also the farmer's crops.

Once your landing site has been decided on, you will then need to set up for the approach. Unless you are flying at a relatively high altitude, you will want to keep the turn from your position to the final approach heading at less than 180 degrees. Using turns of greater than 180 degrees of heading change can result in excessive altitude loss and make it more difficult to reach your intended landing area. It will also be more difficult to judge your touchdown point if you are executing turns of greater than 180 degrees. If you will recall our discussion on determining your glide in relation to the intended point of landing, you will need to judge whether your emergency landing zone is moving up or down the windscreen or is fixed in its position. Turns of less than 180 degrees from the point you lose the engine to final approach heading help make it easier for you to watch the movement of the landing area on the windscreen. Figure 8-3 shows how you

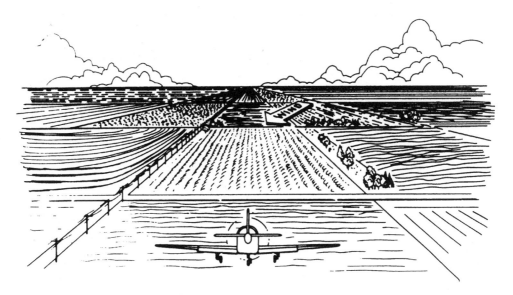

Fig. 8-2 *Landing parallel to field furrows/wind direction.*

can use a 90-degree turn from several points on a base leg to help land at the right spot.

I have taught emergency landings to quite a few students, and nothing is more distressing for them than to set up for the landing area they have picked out, only to find that they are going to land short of the field or that they are so high they will overshoot it. Knowing how to judge your airplane's glide characteristics and hit your intended landing zone takes practice. You can see how important it is to fly consistent approaches every time you land so that when you are faced with a true emergency landing, you will know how much gliding distance you have and whether you can hit the field you are shooting for. In some cases you can compensate for being too high by slipping the airplane with a forward slip. This presents the side of the fuselage to generate more drag and will help you lose altitude without excessive increases in airspeed. Extending flaps may also be used to help get rid of excess altitude if you are too high. Only add flaps according to the manufacturers' best glide recommendations until you are certain you will be able to make the field. It is always better to come in a little on the high side, then bleed off the excess altitude to get down. It is much more difficult to get altitude back without the aid of the engine.

So far we have established the best glide speed and picked our emergency landing area as the first two steps we will perform after losing an engine. Now let's take a look at the other steps you will want to complete as you continue the glide to the emergency landing field. Your plane should be equipped with an emergency checklist that you will follow to attempt to restart the engine if time and altitude permits, and then to safe the airplane prior to landing. This example is for reference.

You should review the emergency-landing checklist in the plane you fly. After the best glide speed had been established and you have picked out the landing area, you will want to quickly determine if you can restart the engine.

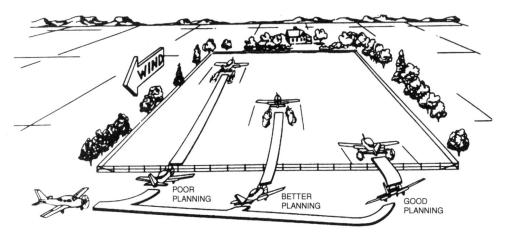

Fig. 8-3 *Emergency landing turn to final.*

Check for proper mixture setting and that carburetor heat is on. In many engine-out situations, carburetor ice has formed, causing the engine to stop. Quickly applying carb heat may melt the ice and allow the engine to restart. Keep in mind that most carb heat is supplied by air flowing through a baffle around the muffler. The air is heated by the engine's muffler and then flows to the carburetor, where it warms the ice. If you wait too long before you apply carb heat, the muffler will cool and may not be able to generate sufficient heat. You will also want to check to see that the fuel selector is on a tank with fuel in it. Since fuel gauges fail, or fuel lines become clogged, change to another tank if one is available. Also turn on the boost pump, if your plane is equipped with one, to help ensure that fuel is reaching the engine. You can also switch the mags from left, right, and both to see if that has any effect on the engine. Incidentally, as long as the propeller is windmilling, the engine should start again without the use of the starter. I used to fly with a Piper Cub pilot that would shut his engine down on purpose, then slow the plane until the propeller quit turning. It was very spooky the first time I saw the prop come to a halt and heard nothing but the wind whispering by the wings. This particular airplane did not have a starter and needed to be hand-propped to start the engine. But when we were ready to restart the engine while in the air, he would just dive the plane until we had enough airspeed to windmill the propeller and the engine would start right up (much to my relief).

All of these steps take a little time to check, but quite often engine loss is caused by one or more of these key items being incorrect. Use your airplane's checklist to make sure you don't miss any of them. You'd be surprised how easy it is to get tunnel vision when the engine stops generating power and miss an important step. Let's say we've checked all of the important steps (fuel, carb heat, mags, etc.) and the engine will not restart. Now we need to get serious about the emergency landing and step through that part of the emergency checklist. Following is an example of typical steps in an emergency-landing checklist:

- Airspeed: Set to best glide.
- Mixture: Idle cutoff.
- Fuel selector: Off.
- Ignition switch: Off.
- Wing flaps: As required.
- Master switch: Off.
- Doors: Unlatched prior to touchdown.
- Touchdown: Soft-field attitude.
- Brakes: Apply as needed.

This checklist is not comprehensive and should not be used in place of the one your plane is equipped with. But it does show the steps that typically need to be taken as you make your approach to landing. The reason you shut down fuel and the master switch is to reduce the possibility for fire. Without fuel or an ignition source, it will be more difficult for a fire to accidentally start. Unlatching the door(s) prior to touchdown helps to keep them from becoming jammed in the closed position if the plane should flip over onto its back, or become bent if the plane strikes something after touchdown.

When landing in extremely rugged terrain, such as a forest or rock area, you will want to touch down at the absolute minimum airspeed to help reduce the chances of injury. In some cases it may be best to actually stall the airplane just above the trees or ground to let it drop just about vertically onto the landing area. This reduces the forward momentum of the plane and helps let the plane settle to the surface as gently as possible. This landing technique is one you probably will not get to practice but has saved the lives of pilots who have used it in real emergencies where the only landing area was on top of a forest of trees. The plane is normally damaged a great deal, but the occupants can often walk away uninjured.

As you can see, emergency landing situations can be dealt with in a controlled manner that improves the chances you will land safely with little or no damage to yourself, your passengers, or the airplane. The key is to keep thinking and to not let the situation get away from you. With altitude on your side, you have time to set up the airplane and pick a favorable landing spot. Even landings in rough terrain can be dealt with if you plan and execute the landing for the situation. Now let's take a look at what you can do to deal with one of the more serious engine-out situations—loss of the engine shortly after takeoff.

Engine Loss after Takeoff

Losing an engine shortly after liftoff can be a very serious situation that demands you act quickly and correctly to execute a forced landing. In this section we will cover some of the considerations associated with this particular emergency situation, from the need to immediately recognize the fact that the engine has lost power, to picking out a suitable landing site within reasonable capabilities of the plane.

Chapter Eight

As you read this section, think about the airports you normally fly from, and the area surrounding them. This will be the type of emergency landing area you will have available to you in the event you should ever be faced with this engine-out scenario. If you are unable to recall what the terrain or placement of open areas around the airports you fly from is like, then you need to make a mental note and survey the area the next time you fly. Knowing potential landing sites before you ever take off can help improve your odds in an emergency during takeoff. You should also give some thought to the altitudes you might need to safely glide to some of those potential landing areas. Under 500 feet of altitude, you are very limited in the landing options you can safely get to. The next time you are climbing out, notice where the plane is as you climb, the altitude, and how close potential open areas for an emergency landing are. You will be surprised at how little altitude 500 feet is, and how short the glide distance you will be able to achieve is. Let's continue with a discussion regarding the need to recognize that you have lost an engine quickly.

The key to successfully making an emergency landing from an engine loss shortly after takeoff is to quickly recognize that the situation exists and react accordingly. When faced with an engine loss shortly after takeoff, the average pilots waste precious seconds in disbelief that the supposedly impossible has happened. You should always consider the possibility of just this situation every time you begin the takeoff roll. What will you do if the engine develops problems before liftoff, immediately after you begin to climb, or at several hundred feet of altitude? By having a plan already in mind and consciously reviewing it before the takeoff, you will not waste time trying to think of what you should do next. A little preparation in this particular case can save those seconds you would spend planning and enable you to use them to control the airplane and head toward a suitable landing area.

If you have lost the engine before takeoff, your only choice is to stop the plane as quickly and safely as possible. Depending on how much runway is in front of you, and how fast you are going, this may require only normal braking action.

Under more severe circumstances, you may not be able to bring the airplane to a halt before leaving the runway. If flaps are extended, you might be able to achieve better braking action by retracting them to place more weight on the landing gear. As you brake, be sure to hold the elevator back enough to prevent the nose of the plane from digging in too much, since the nosewheel will have a tendency to dig into the ground as you leave the runway if there is a great deal of weight transferred to it due to braking.

You may also find that there are obstructions such as landing lights off the end of the runway that could cause serious damage to the plane. If this is the case, pick where you want the airplane to go and head that way. Don't focus on just one spot, such as just in front of the airplane; slightly to the left and right of the runway may present less cluttered areas that would cause less damage to the plane as it leaves the runway. While there is no one right answer, you should try to keep options available. Here again, planning ahead can help you make the best decision before trouble ever starts. There will be limitations on how much you can turn the plane from an aborted takeoff, so be sure that the options you decide on ahead of time fit within the plane's capabilities. If you have time before you leave the runway, turn off the master switch and fuel to avoid potential fire hazards in the event the plane flips over or strikes something during the rollout. There will normally be little time to react to an engine loss before takeoff, so planning ahead can make a significant difference.

If you lose the engine just after takeoff, your options are very limited. The first thing you should do after losing the engine is to get the nose down and set up for best glide airspeed. Keeping the nose in a climb attitude as you quickly scan for potential landing sites will result in a loss of airspeed and a potential stall/spin. Remember that the first rule in any emergency is to fly the airplane. If the runway ahead of you is sufficient, you might be able to land the airplane and safely stop it on the runway. If the standard grove of trees or building is located at the end of the runway, you have fewer options available. From a low altitude you may only be able to change the airplane's heading five to fifteen degrees to the left or right of the runway's heading, limiting your landing options. Don't be tempted to try to bank too steeply in an attempt to make it to a potential landing area. This will cause faster altitude loss and could result in the wingtip of the plane digging in, potentially causing the airplane to cartwheel after the initial impact.

There is no doubt that losing the engine at 50 to 100 feet of altitude is a very serious situation. Pick the best open spot available and set up to land there. Set up for the glide and go through the emergency landing checklist as quickly as possible. If you have time, switch fuel tanks and try the boost pump in an attempt to restart the engine. As you set up for the landing, do your best to avoid potentially dangerous objects on the ground and land at as slow an airspeed as possible. Brake appropriately to safely bring the airplane to a halt.

With greater altitude, your landing site options become greater, but once again you should not attempt to make excessively steep banks or try to return to the runway for a landing unless you have sufficient altitude to safely execute the turn and landing. The FAA has published a document titled "The Impossible Turn," document number FAA-P-8740, which discusses making emergency landings from an engine loss during takeoff. I highly recommend that you obtain this from your local FBO or FAA office and read it. There is an example in this document that gives some sobering information regarding attempting to turn back to the runway for a landing. The first example is for an airplane that has lost its engine at 300 feet altitude. In order to make it back to the runway, you will need to determine how tight the turn must be. Steeper banks will result in a smaller turn radius but greater altitude loss. For this example, let's use a standard rate turn at 70 knots airspeed resulting in a turn radius of 2240 feet. In addition to the 180 degrees of turn you will need to reverse direction, you will also need to turn an additional 45 degrees to turn back toward the runway. Assuming it takes you the average 4 seconds to initially react to losing the engine, then 60 seconds for the first 180 degrees of turn, plus an additional 15 seconds for the next 45 degrees of turn, you are now 79 seconds into the engine loss. If you are losing altitude at 1000 feet per minute during this time, you will lose a total of 1316 feet. Since we started at 300 feet, this poses a serious problem.

Now let's look at the same situation, but with a tighter turn to help us get around more quickly to the runway. Since the bank is going to be tighter, our stall speed will increase. This means we need to maintain a higher airspeed to avoid an accidental stall and can achieve this additional airspeed only by lowering the nose of the plane to gain this speed. Now we have a safe airspeed, but we have also increased our rate of descent, which poses additional problems. In example 2 we can see that a 45-degree bank reduces the turn radius to 560 feet and requires only an additional 10 degrees beyond the 180-degree turn to achieve the runway. The time needed to complete this scenario is the 4-second reaction time, plus 15 seconds for the 180-degree portion of the turn and an additional second for the last 10 degrees of turn for a total of 20 seconds. Let's not even take into account the increased rate of descent this turn situation would present, but use the 1000-foot-per-minute altitude loss in the previous example. Even in this case we will lose 333 feet of altitude, again greater than the 300 of altitude we started with. Before you begin to think that it's just 30 feet or so, keep in mind that the plane is still in a 45-degree bank and will require time and altitude to roll out and get into a safe landing attitude. Cartwheeling next to a runway in an attempt to make it back to the airport is no safer than anywhere else. This is where sound judgment becomes extremely important in order to safely get the plane down.

You should seriously consider whether you can safely get your airplane turned around and back to the runway before you ever make the attempt in an emergency situation. Take the plane to a safe altitude and test the turn radius and altitude loss at different bank angles. You will be surprised at how much altitude you actually lose in these practice sessions. The FAA recommends that you have at least 600 feet of altitude before you attempt to make a turn back to the runway you took off from. To make it back to the airport, you will need to set up a 9-degree-per-second turn, achieve the best glide speed right away, and add 10 knots of airspeed for safety from stalls. The FAA document recommends that this should be done only when confronted with no other options for landing safely. Every airplane will react differently, and some will need more altitude while others may need less. The point of this discussion is that you should be aware of the reality of how much altitude your plane can lose in an attempt to return to the runway after an engine loss.

What are your options if you are faced with an engine loss from an altitude that does not let you safely return to the runway for landing? Once again, this scenario is discussed in the "Impossible Turn" document. It recommends that you scan the terrain ahead of you 60 degrees to the left and right of your initial heading for potential safe landing sites. Figure 8-4 illustrates this concept. You can see that a lake is situated ahead and to the right of the plane's heading, with homes slightly to the center and left. Further to the left is an open landing area that could safely be used by the airplane for an emergency landing. By altering your course to the left, the airplane has a suitable landing zone. Below 300 feet of altitude, you may need to narrow the scan to less than 60 degrees left and right of your heading, but this will depend on the airplane's glide characteristics. The emergency landing site must also be within gliding distance of the airplane from the altitude you are at.

During an emergency landing you should follow the procedures in the operations manual for the plane you fly. However, there are a number of general steps you should normally take that include:

1. Set the pitch angle to achieve the best glide airspeed. This will result in the greatest glide distance for a given altitude.

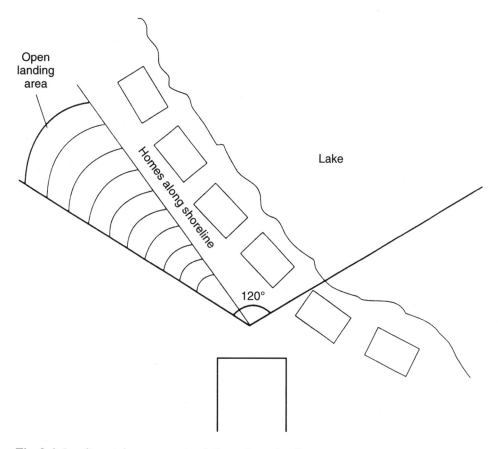

Fig. 8-4 *Landing sight scan: engine failure after takeoff.*

2. Scan for suitable emergency landing sites within the 120-degree arc we previously discussed.

3. Lean the mixture to idle cutoff and turn off the fuel and magnetos to reduce the potential for fire hazards.

4. Use shallow turns to avoid obstacles. Steep turns are more dangerous at low altitudes and can result in stalls, spins, or digging in a wing tip.

5. Lower your flaps after you are absolutely certain you will reach your landing site. Have full flaps extended prior to touchdown and use the airspeed recommended by the airplane's manufacturer.

6. Prior to touchdown, turn off the master switch and unlatch the cabin doors to prevent them from jamming during the landing.

7. If the plane is equipped with retractable gear, use the manufacturer's recommended gear position for emergency landings. If the terrain is rough and likely to result in the gear collapsing, you should keep the landing gear

retracted, landing on the plane's belly. This will help prevent the plane from cartwheeling or flipping if the gear should collapse after touchdown.

We will end this section on emergency landings by restating two very important components of making a safe emergency landing: Fly the airplane and keep thinking. While it may seem redundant to keep repeating these ideas, they are the keys to safely flying an airplane, whether you are faced with a real emergency or just making another flight. Unfortunately, too many pilots fail to do both of these things when a situation turns bad, and they lose not only control of the situation, but also the ability to regain control. Think about emergencies that you may be faced with and plan for them. None of us want to believe we will ever be in a true emergency situation when we fly. We want to fly for the pleasure and sense of accomplishment flying brings to us. But by planning ahead, we can effectively work through those unusual situations life tends to throw at us now and again and be ready to fly another day. And who wouldn't feel pride in being able to lose an engine, safely land the plane, and walk away? Having been through that situation myself, I can tell you it feels good to look back at the plane, knowing that you didn't panic and you safely worked through the emergency.

ACCIDENT AVOIDANCE

We have discussed accident avoidance in many different forms so far in the book, but at this point we are going to take a closer look at what can set you up for accident situations, and how to avoid them. A large portion of this discussion will involve what is known as the *poor judgment (PJ) chain*, which is an interesting way of saying how you make decisions. No pilot intentionally sets himself or herself up for an accident or makes poor decisions with that intent in mind. But accidents do happen, and there are ways to minimize your chances of ending up in one. The PJ chain plays a role in these situations and for that reason we will include it in our discussion.

FAA publication FAA-P-8740-53, "Introduction to Pilot Judgment," discusses the poor judgment chain in great detail. We have covered some of the concepts contained in the document throughout the book so far, but essentially the FAA document states that most accidents are the result of a series of events. The example used in this document is a pilot that is noninstrument rated, has a schedule restraint, and is running late. This pilot has limited adverse weather experience but decides to fly through an area of possible thunderstorms at dusk. Due to his lack of instrument experience, combined with the darkness, turbulence, and heavy clouds, he becomes disoriented and loses control of the airplane.

As you read that little example, you probably identified several poor decisions the pilot made that got him into trouble. Time was a big motivator for the pilot and his decision to press on when the weather conditions were questionable. He was behind schedule and had the classic case of "get-home-itis" that afflicts every pilot from time to time. He also had minimal instrument experience and made a conscious decision to fly into a known thunderstorm area at dusk. The pilot had several opportunities to avoid the accident that resulted from the series of events and decisions that were made along the way.

First, the pilot could have just decided to wait until conditions improved and the flight could have been made safely. This would have necessitated that the pilot overcome the pressure of meeting the schedule, but it is always better to arrive late than to never arrive. Second, the pilot could have altered the route of flight in order to fly around the area of thunderstorms. Like waiting, this would have resulted in arriving later than desired, but at least they would have arrived. Once the pilot encountered weather, he or she could have altered course to get out of that area, either turning back or in a direction away from the storm. Too often, though, pilots "lock on" to an idea and are hard pressed to consider other options as an alternative to the one they have decided on. Finally, once the pilot encountered inclement weather, he should have trusted the instruments and maintained control of the plane, as opposed to becoming disoriented and losing control of the aircraft (Introduction to Pilot Judgment," p. 1).

Two major principles play a role in the poor judgment chain: Poor judgment increases the probability that another will follow, and judgments are based on information pilots have about themselves, the aircraft, and the environment. Pilots are less likely to make poor judgments if this information is accurate. Essentially, this means that one poor judgment increases the availability of false information, which might then negatively influence judgments that follow ("Introduction to Pilot Judgment", p. 3).

As a pilot continues further into a chain of bad judgments, alternatives for safe flight decrease. One bad decision might prevent other options that were available at that point from being open in the future. For example, if a pilot makes a poor judgment and flies into hazardous weather, the option to circumnavigate the weather is automatically lost. ("Introduction to Pilot Judgment," p. 3). By interrupting the poor judgment chain early in the decision-making process, the pilot has more options available for a safe flight. Through delaying making a good decision, the pilot may reach a point at which there are no good alternatives available. Get into the habit of making the best decision you can based on the information available, and do not be afraid to change your mind as additional information becomes known.

Three Mental Processes of Safe Flight

The same FAA document also covers three mental processes related to safe flight. These processes include automatic reaction, problem resolving, and repeated reviewing. Good pilots are actually performing many activities at the same time while they fly. Altitude, heading, and attitude are all constantly monitored and the airplane is adjusted to maintain the desired values. After a period of time, pilots no longer think consciously about what they need to do with their hands and feet to make the airplane do something; it just happens as they automatically make the controls move in the proper way to fly the plane the way they want. This is known as *automatic reaction*.

Problem solving is a three-step process that includes:

1. Uncover, define, and analyze the problem.

2. Consider the methods and outcomes of possible solutions.

3. Apply the selected solution to the best of your ability.

Through taking these steps, you will improve your ability to understand problems and resolve them. By correctly determining the actual cause of a problem, rather than misunderstanding it, you can aid in making better decisions to resolve it. The poor judgment chain can be avoided or broken through the use of good problem-solving skills.

The last mental process is repeated reviewing. This is the process of "continuously trying to find or anticipate situations which might require problem resolving or automatic reaction." Part of this skill includes using feedback related to poor decisions. As you fly, you need to constantly be aware of the factors that affect your flight, including yourself, the plane, weather, and anything else that could be a factor. Through remaining "situationally aware," you will be better informed and able to analyze the actual conditions you are flying in ("Introduction to Pilot Judgment," p. 5).

Five Hazardous Attitudes

The last topics in the "Introduction to Pilot Judgment" we will cover are five attitudes that can be hazardous to a safe flight. These attitudes include:

1. Anti-authority: "Don't tell me!"

2. Impulsivity: "Do something—quickly."

3. Invulnerability: "It won't happen to me."

4. Macho: "I can do it."

5. Resignation: "What's the use?"

Each of these attitudes can get in the way of a pilot making a good decision. Over-inflated ego, lack of confidence in their abilities, and the need to prove themselves are just a few personality traits that pilots can exhibit that can cloud their judgment. This poor judgment mindset is not because these pilots intend to make bad decisions, but because they are influenced by behavioral traits that negatively influence their ability to make a good decision. We all have these traits to one degree or another, but how much we let them influence us is a big factor in whether we can see through them to make a good decision. While you are monitoring the plane, weather, and other aspects of your flight, also monitor yourself. If you find that any of these traits are getting in the way of a good decision, it's time to take a step back and rethink the situation.

Accident Prevention

The poor judgment chain we just covered is a big influence in how many pilots get into accident situations. Through poor planning and an inability to recognize the need to reevaluate a situation, pilots can put themselves in harm's way. We know we need to use sound decision-making guidelines, but exactly what does this mean? In this section we will cover some of the steps you can take to avoid accidents or emergency situations. While this is not all encompassing, the ideas in this section should give you a foundation to build on as you begin to think about some of the habits you may have developed.

First, think about the risks associated with an action before you take it. Take two pilots who decide to fly under a railroad bridge over a river. The first hops in the plane, heads out to the bridge, and flies underneath it. The second takes a boat out to the bridge, inspects it for wires that may run under it, measures its height and width, and sees how the terrain is around the area. He then goes back and measures the plane's dimensions, checks the winds, and then when he feels it's safe, goes and flies under the bridge. Both pilots executed the same maneuver, but the second pilot planned his actions, and waited until he knew it was safe to fly. The first showed very poor judgment in flying under the bridge because he did not have all the facts.

How many pilots don't have all the facts when they fly? Are they within weight and balance limits? How much fuel will they actually burn during the flight? What is the weather along the route of flight? The list of the factors that a pilot should understand is very large, but manageable. Before you fly, understand what it is you want to do, decide what the limiting criteria for the flight are, and adhere to them. Whether the flight is a hop around the patch or a cross-country from New York to Los Angeles, you need to determine what you want to accomplish. This may seem like common-sense stuff, but not enough pilots take advantage of all the information available to them and weigh the results of their actions.

Use your checklists when you fly. We all forget things, and using a checklist will help us avoid making embarrassing, if not serious, errors in operating the airplane. Every retractable-gear airplane has a checklist with some reference to the gear being down before landing, yet every year pilots manage to land with the gear up. We become comfortable with a plane that we fly on a regular basis, and it is not uncommon for anyone to become somewhat complacent. Checklists can help us avoid missing something we need to do as we take off, cruise at altitude, or land.

You should also know the airplane you are flying. As much as familiarity can generate complacency, lack of experience with a plane can get a pilot into trouble. If you are learning to fly an aircraft that is new to you, read the operations manual. Learn the airspeeds, proper engine operation techniques, and every other bit of information regarding the plane. General Chuck Yeager once stated that knowing the systems of the test aircraft he flew gave him the ability to recover from situations other pilots may not have been able to. The same holds true for all of us. Learn about the fuel system, how the emergency gear extension works, and everything else you can read about. Poke around the airplane to verify how systems work. This knowledge could be very useful in an emergency situation, or in just making you a more competent pilot.

Finally, know your limitations. Many accidents are caused because pilots push the airplane or themselves beyond what they are capable of. If you are not comfortable with your proficiency in a certain area, get instruction from a qualified flight instructor to gain experience and confidence. Do you feel good about your crosswind landing techniques, or does the plane tend to stray across the runway as you land? How about stalls? I fly with quite a few pilots who are not really interested in aerobatics as much as how to recover from a full stall or an unusual attitude. Most of us fly in a routine manner, to a small number of airports that we become accustomed to. Are you comfortable flying into a runway that has a real obstacle at the end of it or is truly a short runway? These are the things that cause accidents. Pilots get into situations that are not really dangerous but are beyond their level of experience.

Realize that when you are tired, sick, or under stress, your mind may not be clearly focused on flying, and you are more prone to making mistakes under these conditions. Don't fall prey to the personality traits we covered in the poor judgment chain sections. If you don't feel good about making a flight, for whatever reason, don't make it. Wait until the conditions become satisfactory before you fly. Real pilots are those who can say they will stay on the ground until conditions become safe.

CONCLUSION

We have covered a number of safety-related topics in this chapter: emergency procedures for a number of different scenarios, including being lost, bad weather, and engine loss. We also discussed the decision-making process and how it relates to avoiding a bad situation or getting out of it once you are in one. No book can cover every aspect of this topic, but we have reviewed a number of concepts that are important to safe flight.

It has been stated many times in the book so far, but safety should always be your primary consideration when you fly. Whether you are taking a VFR joyride or climbing into the space shuttle for a trip into orbit, understand the safety factors associated with the conditions of the flight, the aircraft you are flying, and yourself. If you get into the habit of thinking in this manner every time you fly, you will improve your proficiency, safety, and pride in how you fly.

9
Basic Instrument Flight

The flying we have discussed to this point in the book has been covered with only VFR (Visual Flight Rules) weather conditions in mind. This means that you have at least one-mile visibility in uncontrolled airspace, to three miles visibility in controlled airspace. There are also cloud clearance restrictions that must be adhered to in order to remain in VFR conditions. In this chapter we will discuss flying the airplane solely by reference to the instruments inside the airplane, or while the plane is operating under IFR (Instrument Flight Rules). This task demands practice, proficiency, and concentration to accurately maintain heading, altitude, and your position as you fly using only the airplane's instruments.

The goal of the chapter is to give you a basic understanding of how to maintain heading, attitude, and altitude as you fly, and the discussion is for a very limited set of emergency conditions. We will cover the basics of what the instruments provide, how to scan them, and how to use the instruments to guide the airplane to safe conditions. In the previous chapter we used the example of a pilot getting caught in IFR flight conditions through improper planning and the desire to press on into a known thunderstorm area. There is always the possibility that you could be caught in unexpected weather conditions that might reduce your visibility to IFR conditions, and you should be able to fly the airplane safely out of those situations by using the instruments.

As you read the chapter keep in mind that this instruction should have been part of your private pilot flight training. But if it has been some time since you have had the opportunity to fly using instruments, you should get some refresher training to gain additional experience. Flying on instruments alone is very different from using them in conjunction with being able to look outside the airplane to cross-reference what your attitude and heading are. While you should never go looking for instrument conditions to fly in as a pilot that is not IFR rated, knowing how to navigate using instruments could be very handy if the unexpected takes place. Let's begin with a look at what the basic instruments are and what they tell us.

INSTRUMENTS

In most general-aviation instruments there are several mechanisms that are normally placed in standard positions relative to each other. With that said, you will probably find that the airplanes you fly have some variation of the instrument panel configuration that we will be using here. Having the instruments placed in a different position in the panel is not a real problem, but know where the instruments are so that you don't waste time looking for them and are able to scan the instrument panel effectively. As you will see, there are normally ways to cross-check instruments using the proper scanning techniques that can help you quickly determine if you have a malfunctioning instrument. Knowing where to look on the panel can help you smoothly scan the panel and more easily maintain the proper attitude, heading, and altitude.

Figure 9-1 shows a basic instrument panel with a series of six instruments laid out in it. Across the upper row from left to right are the airspeed indicator, the artificial horizon, and the altimeter. On the second row left to right you will find the turn and bank indicator, the directional gyro, and the vertical speed indicator. Each of these instruments has a significant piece of information to offer in reference to the airplane's status. Let's work our way across the panel and look at each instrument's function and relation to the other instruments.

Fig. 9-1 *Instrument layout.*

Airspeed Indicator

The airspeed indicator is a pressure-sensitive instrument that uses pressure differences to determine the speed of the airplane. The pitot tube and static system measure the pressure of air and use the differences between them to calculate the airspeed. As the airplane flies through the air, the pitot tube measures the pressure of air entering the pitot tube, which is placed along the wing on most general-aviation aircraft. Figure 9-2 depicts a typical static system. As you can see, the pitot tube connects to the airspeed indicator, which then connects to the static vent and alternate static vent. The altimeter and vertical speed indicator are also connected to the static system. The pitot is positioned so that the airflow impacts it and flows into it, generating pressure. As we have already discussed in a previous chapter, the airspeed indicator is marked with several colored arcs that show ranges for normal flight, flap extension, stall speeds, and the never-exceed speed.

But you can also use the airspeed indicator to cross-reference with the artificial horizon and vertical speed indicator to verify that they are also working correctly. If you are in a dive, you will notice that the airspeed will increase. In a climb it will decrease as the nose rises. If you keep power settings constant, a drop in airspeed could mean the nose is rising. Cross-referencing this information with the artificial horizon, altimeter, and vertical speed indicator will help you determine if the instrument is accurate. If the other gauges give an indication that the plane is in level flight but the airspeed indicator is decreasing or increasing, it could be a sign of a malfunctioning airspeed indicator.

While flying under instrument conditions, the airspeed indicator is used to assure that in climbs or descents the plane's airspeeds are maintained within the correct ranges. During a climb or descent, the attitude indicator can be used to control the pitch, but the airspeed indicator is used to refine the airspeeds to the correct values. In a climb you want to use the correct airspeed for the situation, such as V_x or V_y, while in a descent you

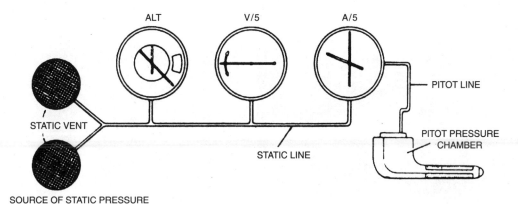

Flush-type static source

Fig. 9-2 *Static system.*

want to also be sure you are not exceeding recommended speeds. When pilots become disoriented under instrument conditions, it is not uncommon for the nose to drop and the airspeed to build rapidly. By monitoring the airspeed indicator, this problem can be more easily avoided.

Artificial Horizon

The artificial horizon (AH) is one of the most useful, and most abused, instruments in the panel. Both bank attitude and pith angles are available from the artificial horizon, which is extremely important when flying on instruments. But many pilots have a tendency to focus almost exclusively on the AH to the point of excluding the other instruments. Illustrated in Figure 9-3, the artificial horizon is normally made up of a "ball" that depicts the sky and ground, the intersection of those two areas being the horizon. While the color of the sky and ground will be different for various AH instruments, they will normally use blue or white to represent the sky, and brown, black, or other darker tones for the ground. Graduated lines shows pitch up and down angles on each of these areas. This central "horizon" portion of the instrument moves up and down to represent pitch angles and rolls left and right to show bank angles. Graduated marks along the sides of the artificial horizon are used to measure the bank angle of the airplane. In the center of the instrument there is usually some representation of the airplane that allows you to more easily visualize the pitch and bank attitudes for the plane.

The artificial horizon is a vacuum-driven gyroscopic instrument on most general-aviation aircraft. Essentially, the vacuum pump on the airplane's engine generates a flow of air through the instrument that spins up a gyroscope that maintains a constant attitude relation to the earth. Figure 9-4 shows a typical pump-driven vacuum system and the instruments normally attached to it. In most general-aviation planes, the directional gyro, artificial horizon, and the turn and bank indicator are driven by the vacuum system. As the plane banks or changes pitch angle, the gyroscope pivots on its mounts, trying to stay level. The central horizon is attached to the gyroscope and also maintains a relative posi-

Fig. 9-3 *Artificial horizon.*

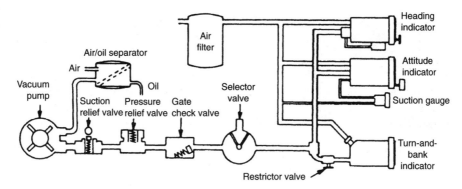

Fig. 9-4 *Pump-driven vacuum system.*

tion to the earth. As the plane changes pitch or bank angles, the gyroscope holds its position, which allows the instrument to show the bank and pitch angles.

The artificial horizon is useful, but it should also be cross-checked against the altimeter, the airspeed indicator, the vertical speed indicator, and the directional gyro. For instance, if the AH shows the plane is in a bank, but the directional gyro shows the heading is constant, there may be a problem with one of the instruments. A further cross-check with the compass will confirm if it is in sync with the directional gyro. If the two match, there is a strong likelihood that the artificial horizon is inaccurate. On the other hand, if the compass is showing a turn and the directional gyro (DG) is not moving, the DG could be malfunctioning. You can begin to see from just this little discussion that flying on instruments demands that you understand the relation of the instruments to each other and know how to rule out specific instruments when one or the other is malfunctioning. It also means that attempting to fly on instruments when you are not rated to could be more demanding than you might have anticipated.

Altimeter

The altimeter is used to determine the altitude of the airplane through the measurement of barometric pressure. The altimeter is adjusted to the correct pressure by setting the value in the small window to the current barometric pressure. The altimeter shown in Figure 9-5 shows an altimeter setting of 30.34. Like the airspeed indicator, it uses the static system as the source for its measurements. Altimeters can be equipped with one, two, or three needles used to indicate altitude. On an altimeter equipped with three needles, the long needle indicates hundreds of feet, the next shorter needle indicates thousands of feet, and the shortest needle shows tens of thousands of feet. If the barometric pressure in the window is incorrect, the altitude the instrument indicates will also be incorrect. Assuming your altimeter is working correctly, you can also determine if you are climbing or diving based on whether it shows your altitude is increasing or decreasing. This can also be useful for cross-checking with the artificial horizon and the vertical speed indicator to determine if all instruments show the same information regarding climbs or descents.

Fig. 9-5 *Altimeter.*

Turn and Bank Indicator

Figure 9-6 shows a turn and bank indicator, also known as a turn and slip indicator. You can see that it is made up of a vertical white bar, called a needle, and a black sphere in a glass tube, called the ball. We have already covered in great detail the fact that the ball is used to indicate whether the plane is in coordinated flight. Figure 9-7A shows the needle and ball when the plane is in a slip, while 9-7B shows the plane in a skid. In both cases the ball is not centered in the tube, indicating that the plane is in uncoordinated flight. The phrase "step on the ball" is a general rule of thumb that says you should step on the rudder to the side that the ball is toward. In 9-7A, this would mean you need to push on the right rudder, while in 9-7B you need more left rudder.

The needle indicates the direction of the bank, which is to the right in both examples. Located above the needle are three figures. The middle white mark above the needle is the centered marker, while on either side the dog-house shaped marks are used to indicate the plane is in a standard rate turn. If the ball is centered and the needle is located on the center mark, the plane is in a wings-level attitude. But if the needle is to the left or right of the center marker, the plane is banked. This can be a useful indicator of the plane's attitude and a good cross-check of the artificial horizon. If the needle is on the left or right "dog-house," and the ball is centered, the plane is making a turn in that direction at 3 degrees per second of heading change. When flying under instrument conditions, turns are normally made at the standard rate. The timing and layout for most instrument approaches are set up for standard rate turns.

Fig. 9-6 *Turn and bank indicator.*

Fig. 9-7A *Turn and bank: slip.*

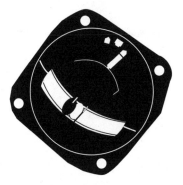

Fig. 9-7B *Turn and bank: skid.*

Like the artificial horizon, the turn and bank indicator is driven by a gyroscope. This gyro is driven by the vacuum system on many aircraft, while on others it is an electrically powered gyroscope. The reason for having the turn and bank on a separate drive system is that if the vacuum pump fails, you will not lose all gyroscopic instruments. In the unfortunate event that the vacuum system does have problems, you could still maintain control of the plane by using the turn and bank indicator—in conjunction with the compass, altimeter, and vertical speed indicator—to maintain level flight at the correct heading. Granted, this requires a little more work on the pilot's part, but the redundancy of instruments gives you a fallback in the event of major system problems.

Directional Gyro

A directional gyro (DG) is depicted in Figure 9-8. As you can see, it has the headings of a compass on it, with a small plane in the center of the instrument. As the name implies, this is a gyroscopically driven instrument that allows you to determine the heading of the plane. The DG is basically a dumb instrument; it does not inherently know magnetic directions. Before you fly you must set the initial heading on the DG to match the compass. After the heading is set, the gyroscope in the DG attempts to maintain its alignment with that initial setting. When the plane turns, the card with the compass headings rotates within the DG, showing the airplane's current directional heading.

Fig. 9-8 *Directional gyro.*

It should quickly be apparent that if the initial heading that the DG is set to is inaccurate, the headings that it shows will also be in error. The gyroscope in the DG also tends to precess, or shift on its axis as time passes. In order to be sure the DG shows an accurate heading, you will need to periodically reset the heading it shows to match the heading on the compass. Depending on the condition of the DG, it may be necessary to reset the heading as often as every ten minutes. The advantage the DG has over the compass is that it will maintain a steady indication of the plane's heading and is not subject to the bobbing and weaving that the compass is prone to in rough air. Anyone who has attempted to hold an accurate heading while flying in turbulent air by using only the compass will attest to the problems this creates. The directional gyro is not affected by this type of movement and allows you to more accurately hold a heading.

When a DG fails, it can slowly precess, the card that shows the compass points slowly turns, or it can fail more completely. In a severe failure of the DG it will often spin around in circles at an amazing rate. Major failures are easy to pick up; the instrument is making so much nonsense that it is very easy to spot the behavior. But in a slow failure the heading can slowly drift off from the correct value. If you do not catch this you might find that you are turning unconsciously as you attempt to hold the correct heading on the DG. While you may think you are heading in the right direction, this situation can put you significantly off the correct one. For this reason you should frequently cross-check the DG with the compass.

Vertical Speed Indicator (VSI)

The vertical speed indicator is the last instrument in the basic cluster we will review. Figure 9-9 shows the instrument, which is another that is attached to the static system. The instrument shows the rate of altitude change in hundreds or thousands of feet per minute. The VSI is useful for setting up descents at the proper rate for a number of instrument approaches. They are often calibrated for 500-feet-per minute descents, which provide a comfortable descent angle and manageable airspeeds during the approach. They are also useful for cross-checking the altimeter and artificial horizon to assure that they are accurate.

BASIC INSTRUMENT SCAN

This section will discuss flying the airplane by reference solely to the instruments we have discussed in this chapter. This is for emergency purposes only, and this is not intended to provide the reader with more than a rudimentary understanding of flying on instruments. Proper planning and preflight checking can prevent the majority of accidental entry into IFR or minimal visibility flight conditions. But for those situations where you encounter weather that requires that you fly by reference to instruments, it is necessary that you understand their function and how to fly properly. In this section we will cover the basics of flying on instruments and the best methods for extricating yourself from the situation.

The first thing you should do when you encounter instrument weather conditions is continue to fly the airplane. Panicking and letting the airplane get into an uncontrolled attitude is one of the fastest ways to lose the ability to salvage the situation. Early in my piloting career, a pilot I knew got into just this situation. The pilot was low time, had only recently received his private pilot rating, and had purchased a high-performance single-engine plane. The pilot took off into marginal weather conditions, encountered IFR visibilities, and lost control of the airplane. The plane impacted the ground not all that far from the airport, killing the pilot. The pilot had likely become disoriented in the clouds, not used the instruments properly, the results being fatal. Remember—keep flying the airplane.

Establish a scan of the instruments as soon as you encounter minimal visibilities. We will discuss the scan further, but maintain the proper flight attitude with the instruments. The second thing you should do is to contact air traffic control (ATC) immediately. While this may seem like something you do not want to do to avoid getting in trouble, that should be the last of your worries in this situation. ATC can help establish your position, direct you to an airport that may be in a more favorable weather situation, and keep you clear of other legal IFR traffic.

Fig. 9-9 *Vertical speed indicator (VSI).*

The primary reference instrument in IFR conditions will be the artificial horizon. This instrument provides both pitch and bank information and will help you maintain the correct attitude. Your scan will move to the directional gyro, the altimeter, and airspeed indicator, essentially working the "T" of instruments this forms on the instrument panel. The turn and bank indicator and the vertical speed indicator should also be checked periodically, but use of the other instruments will help you hold the desired attitude and heading. Avoid the tendency to focus on one instrument exclusively, normally the attitude indicator. Keep the scan smooth and constant. This will help you avoid blocking out important information regarding the airplane's flight status.

Once you have established contact with ATC and have the scan working and the plane in the desired flight attitude, relax as much as possible. Students learning to fly the plane on instruments often establish a "death grip" on the control yoke. This prevents them from feeling the airplane and making smooth inputs on the controls. When flying on instruments, you need to make a conscious effort to use small, smooth control inputs when making changes in bank, pitch, or direction. Pilots often fall into the trap of using too much control input for a given change, then overcompensating once they realize they have gone too far. It is not at all uncommon for pilots to bank well past 30 degrees, then realize as the plane reaches 45 degrees of bank that they are losing altitude, the airspeed is increasing, and they have flown past the heading they wanted to establish. This sets them up for overcontrolling in an attempt to get back to where they want to be. If you keep the control inputs small and scan the gauges in the "T" rhythmically, you will notice the results of the control inputs and be able to make any needed corrections quickly and efficiently.

CLIMBS BY REFERENCE TO INSTRUMENTS

ATC may have you initially climb after you contact them to reduce the chances of encountering hazards on the ground. If that is the case, you will want to use the same climb power settings and pitch angles that you would during a VFR climb. The climb should be made at a constant airspeed, which means that you need to maintain a constant pitch. To begin the climb, add power to climb settings, then use the artificial horizon to establish a gradual nose-up pitch. For most airplanes, a 10-degree pitch-up angle from level on the artificial horizon would be a good place to start. Watch the airspeed as the nose pitches up, and if the airspeed drops below the correct value, slightly reduce the backpressure on the control yoke. Keep the inputs small to avoid overcontrolling the airplane. Once the airspeed is established at the correct climb value, hold a constant pitch angle to maintain the airspeed. You will also need to monitor the altimeter and directional gyro to make sure you do not overshoot the desired altitude or stray from the heading ATC has assigned you. Don't forget to keep the ball centered through proper rudder use during the climb. As you approach the assigned altitude, slowly lower the nose to a level attitude by using the artificial horizon. As the speed builds, reduce power to cruise settings and trim for level flight at that altitude. Keep scanning the gauges; don't focus on just the artificial horizon or the altimeter.

STRAIGHT AND LEVEL FLIGHT AND TURNS

In order to maintain straight and level flight, you should establish the correct power settings, then trim the plane to hold a constant altitude with neutral pressure on the elevator control. Proper use of trim can help reduce your workload and fatigue factor, which is very important in this situation. Use the artificial horizon as the primary indicator of your pitch and bank attitudes, but also scan the directional gyro for proper heading, and check the altimeter for the altitude. The turn and bank indicator is useful when you need to turn the airplane and should be used to establish coordinated, standard-rate turns. If you use standard-rate turns whenever you change headings, you will avoid becoming so steeply banked that you lose orientation. Use of standard-rate turns will also reduce the chances that your turning rate is so fast that you turn past the heading you are turning to. If you do make turns, remember to use a slight amount of backpressure on the control yoke to maintain a constant altitude during turn. It is quite common for new instrument students to overbank the plane during turns and allow the nose to drop as they enter the turn. This allows loss of altitude and can cause a significant increase in the airspeed. It cannot be stressed enough. Keep the control inputs small and smooth, and scan the gauges.

DESCENT BY REFERENCE TO INSTRUMENTS

If you need to descend on instruments, you will once again use the artificial horizon to initially set up the descent angle. If necessary, reduce the engine power setting to avoid gaining too much airspeed. Use the vertical speed indicator in conjunction with the airspeed indicator to establish a safe rate of descent. If you notice the airspeed is becoming too high, raise the nose of the plane and/or reduce power. If you use approximately a 500-foot-per-minute descent rate and power settings at the bottom end of the green arc, you should avoid excessive airspeeds. However, every plane is different, and it may be necessary to use different techniques for the plane you fly.

Like the climb, a descent should be at a constant airspeed and rate of descent. Using a constant pitch angle and power settings is necessary to establish the descent. You should decide on the altitude you want to descend to before you start the descent, then begin raising the nose to a level attitude and increasing power as you approach the target altitude. If you are setting up for landing, it can be a good idea to get the airplane into the proper landing configuration before you begin the descent to reduce the workload during the descent. Continue to scan the airspeed indicator, artificial horizon, altimeter, and directional gyro during the descent. Again, it is common for pilots to focus on just one instrument during descents. This can vary from the artificial horizon to the altimeter, but if you exclude the other instruments from your scan you will not be in complete control of the plane.

INSTRUMENT NAVIGATION

If you become immersed in instrument conditions, descending until you can see the ground and navigating via VFR procedures may not be an option. In severe weather cases, the clouds, rain, snow, or fog may completely obscure the surface and not allow you to descend. The same holds true for executing a 180-degree turn to get out of the weather, or climbing above it. In those instances ATC may direct you to navigate via instruments to an area of better visibility or to an airport that you can land at. This section will cover the basics of navigating by instruments. Like the previous section, this is not intended to be a self-taught instrument course, but information that may be useful in an emergency. You should find a qualified flight instructor to give you proper training in instrument flight, whether you intend to go on to get your instrument rating or are just interested in practicing flying by reference to instruments for proficiency.

VHF Omnidirectional Range (VOR)

The most common instrument used in radio navigation is the VOR. VOR ground stations transmit signals to the VOR receiver in your airplane. The receiver can differentiate and indicate the relationship of the plane from the station through the use of radials. The ground station essentially has 360 radials emanating from it that correspond to the points on a compass heading. Resembling the spokes on a bicycle rim, the receiver in the airplane can determine which radial the plane is over and whether the plane is traveling to or from the ground station. The VOR instrument that is located in the airplane's instrument panel is depicted in Figure 9-10. You can see the "card," which has markings for each of the 360 radials it can detect. You will also notice TO and FROM markings on the right side of the instrument, in addition to a vertical needle at the center of the face. The knob on the lower left marked "OBS," Omni Bearing Selector, is used to dial in the correct radial, the "card" containing the radial marks rotating as the OBS knob is turned.

The VOR receiver is also made up of a receiver that is normally located in the same radio unit as the communications radio. VOR ground stations have frequencies assigned

Fig. 9-10 *VOR instrument.*

to them. In order to navigate using a VOR, you must find the correct frequency for that station, dial that into the VOR radio, then listen for the Morse code identifier. This will positively establish that you have dialed in the correct frequency for the VOR you are looking for. The correct frequency and Morse code identifier are located on sectional charts and allow you to quickly determine the correct frequency for the ground station and the identifier sequence.

To find out which radial the plane is on from a VOR ground station, you will turn the OBS knob on the instrument until the needle centers with the flag located in the FROM location. This can be useful in determining your location. For instance, if you use two different VOR ground stations and locate the radials you are on from them, the intersection of those lines is your location. The easiest method to navigate to a VOR station is very straightforward. Use the OBS to center the needle with the flag located in the TO indicator. This is the approximate heading you will need to get to the station under no-wind conditions. In the VOR illustration, the heading 330 radial centers the needle, so we would turn to approximately 330 degrees heading. The idea is to keep the needle centered as you fly toward the station. Figure 9-11 shows a plane tracking inbound to a VOR station and the effects of wind on the plane's course. If wind is pushing you away from the desired radial, you will end up crabbing into the wind in order to establish the ground track that will correspond to the radial you are flying. The wind correction techniques we used for ground reference maneuvers are just as useful when we are flying a VOR radial to the ground station. As you can see, the pilot corrects for the wind and continues to fly toward the VOR station.

If you are drifting from the radial you have dialed in, the needle will move from the centered position. To correct for any drift, you will correct your heading in the direction of the needle. "Fly toward the needle," is a common phrase when flying inbound to a VOR with the TO flag indication, or outbound from a VOR with the FROM flag indication. In other words, if you are drifting to the right of the desired radial, the needle will move left in the instrument. To correct for this drift you will adjust your heading by turning to the left. You should keep your heading corrections relatively small until you have a good idea how much crab angle will be necessary to maintain your track on the VOR radial. New instrument pilots often overcorrect, then end up chasing the needle back and forth across the instrument.

As you fly over the station and head outbound on the radial, the TO/FROM flag will flip from the TO position to the FROM position. This is useful in determining station passage and your location. As you fly outbound from the station, you can continue to "fly the needle" to stay on the radial. Depending on the situation, ATC may have you fly via several VORs to your destination, but the process is basically the same to navigate to each of them. One thing you should note—if the flag is in the FROM position, but you are actually flying to the VOR, the needle will act in reverse to the action seen when it is in the TO position. This means that you would fly away from the needle to correct the needle back to the centered position. Under most circumstances you will not want to set the VOR up to act in this manner, but if things don't seem to be working correctly, verify that the TO/FROM flag actually matches whether you are flying to or from the VOR ground station.

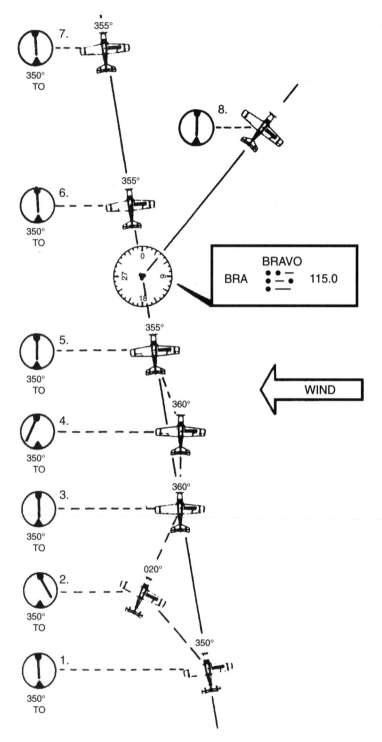

Fig. 9-11 *VOR tracking.*

ADDITIONAL RADIO NAVIGATION AIDS

There are several other radio navigation aids available to a pilot. We will briefly cover those in this section, but the intent is only to make you aware of their existence and basic function. The aids we will cover include Radar Services and the Global Positioning System (GPS). Radar services are commonly used in both VFR and IFR flight and have been in service for several decades at many tower-controlled airports. Additionally, ATC uses radar to track the position of aircraft along routes of flight, providing heading and altitude guidance to pilots. GPS navigation has recently been made available to the general public and is radio navigation provided through the acquisition of data from satellites in orbit around the earth. Originally developed for the military, a small receiver can detect the transmissions from these satellites and establish the aircraft's position with a great deal of accuracy.

If you encounter IFR weather, Radar Services can be of immediate aid in providing heading and altitude directions to you. Center frequencies can be obtained from a variety of sources, but the easiest way to get help quickly is to use the emergency VHF frequency of 121.5 MHz. This frequency is monitored by Flight Service Stations and radar facilities, and the personnel will be able to quickly help you in the event of an emergency. When radar services are requested, the radar operator will initially ask the pilot to set a unique code into the plane's transponder. The transponder is a device that transmits and receives information with the radar facility and enhances the image of the plane on the radar operator's screen. This unique code identifies your airplane on the radarscope, helping the operator to quickly determine where you are. In the event your airplane is not equipped with a transponder, the operator may have you fly a series of turns until they can determine which plane is yours. The operator will also ask you a series of questions that include whether you are in VFR or IFR conditions, the amount of fuel you have, the number of people on board the aircraft, and if you are instrument rated. This provides the radar operator with information that can help them formulate the most efficient plan for getting you safely to an airport.

You should make it a point to occasionally use radar services even in VFR flight conditions. This will help you become familiar with the procedures and abilities of the service and teach you to communicate with them in an efficient manner. They can provide you with traffic information and help you avoid other aircraft. In the event of a true emergency, having this experience will allow you to quickly take advantage of the help they can provide.

The Global Positioning System receiver is a device that has enjoyed widespread use in recent years. At one time hailed as the end to the VOR system, the GPS provides an incredible navigation aid to pilots, boaters, the military and just about anyone that wants to be able to find out where they are and the direction to head to get somewhere else. There are a number of GPS vendors that provide products with a wide variety of functionality. Basic aviation models will allow you to plot a course from point to point, then act somewhat like a VOR, letting you know if you are left or right of the desired course.

More advanced versions of the GPS include what is known as a moving map. This can be anything from a rough line map on a small screen to models that incorporate laptop computers with color maps that resemble sectional charts. In accidental incursions into IFR or low-visibility weather, many GPS systems have a NEAREST function. This takes your current position and determines the nearest airport to where you are. The direction, distance, and flying time are normally provided by the GPS unit in these situations. Many pilots use hand-held GPS units, while some aircraft have them mounted into the panel and tied into the autopilot. While it is not likely to be the demise of the VOR system, GPS receivers provide a tremendous amount of information to pilots and can be a worthwhile investment for any pilot that flies cross-country on a regular basis.

CONCLUSION

Instrument flight and navigation have allowed us to fly in conditions that grounded planes in the early days of aviation. Those early aviation pioneers realized very early on that in order to be a reliable mode of transportation, airplanes would need to be able to fly through clouds and still reach their destination safely. The NDB, Non Directional Beacon, was an early radio navigation aid that allowed pilots to fly to an NDB station, eliminating the need for pilots to navigate by ground references. The artificial horizon and other gyroscopic instruments gave pilots the ability to safely control the airplane in adverse weather conditions.

While we have this technology available to us, the use comes with the need for training and proficiency. It is with good reason that the FAA sets stringent requirements for instrument flight currency; it is very easy to become disoriented when flying through clouds. Even experienced, IFR-rated pilots occasionally get a mild case of vertigo, the inner ear telling them something different than the instruments. By focusing on the instruments and believing what they are saying, the vertigo feelings normally pass and things settle down to a normal flight environment. But it is for these reasons you should not take flying into poor weather conditions lightly.

A VFR pilot caught in instrument conditions may quickly find that without proper training they are in over their heads and uncertain of the next step to take. To avoid trouble, check weather before you fly and after you get into the air. Don't let the desire to get somewhere override good judgment and put you in a precarious position. If you do get into poor visibility, fly the airplane and keep thinking. Contact ATC to help you find your way to a safe area or to an airport you can land at. Even if you never intend to get your instrument rating, I highly recommend that you get instrument dual instruction from a qualified instructor. Knowing how to scan the instruments, maintain control of the plane, and navigate via the radio could be an extremely valuable skill in a tight situation.

10
Conclusion

We've covered a great deal of information in this book. Now that you've completed it you should have a strong understanding of the basics of flying, from how to fly straight and level, to ground reference maneuvers and takeoffs and landings. We also discussed more advanced flight maneuvers such as stalls and spins, chandelles, and lazy eights. Throughout the book the intent has been to give you a practical understanding of flying and to build on the previous information with each step forward in the book. But no one book, or series of books, can capture all of the knowledge that you will require to become a proficient, safe pilot. To achieve that goal you will need to fly, to focus on methods of improving your flying skills, and to obtain continuing instruction from flight instructors that you trust.

In this final chapter of the book, we will cover a few loose ends, items that did not fit into other chapters but still need to be discussed. We will also touch on topics that we did cover but are worth mentioning a final time. Let's start with flight training.

FLIGHT TRAINING

Whether you are thinking about learning how to fly or have 20 years and 10,000 hours of time in your logbook, there is always the need to get good flight training. When you receive flight instruction, it should be from a competent, qualified flight instructor that you feel comfortable flying with. There needs to be a channel of communication between you and your flight instructor that flows in both directions. As you search for a flight school or an instructor, you should define what you want to learn and the qualities that you feel are important in an instructor. We have become a service-oriented society, and you should expect nothing less from your instructors than someone who wants to understand your goals and help you achieve them.

As you first meet your flight instructors, form an impression of them. Do they act professionally, schedule your appointment to fly, then keep that appointment? Do they communicate well, in a manner that is clear to you? Any flight, whether a first lesson or a biennial flight review, should involve pre- and post-flight discussions. The time you spend in the air should be for flying; your instructor should spend time on the ground before the flight explaining what you will cover during the flight. Remember that you have the right to change flight instructors if they are not training you in a way that seems beneficial to you.

For those of you who like to "surf the Web," recently there was a "thread" in the aviation news group regarding a pilot getting checked out in a taildragger. This pilot described the lessons he was receiving and the number of hours it was taking for the instructor to sign him off. It was very apparent that the student was not comfortable with the training he was receiving. He questioned whether the instructor was milking the situation and what he should do. Finally, after nearly fifty hours of dual, the pilot switched to another flight instructor and was signed off for taildraggers soon thereafter. This appears to be a flagrant example of an instructor stretching out the flying time to generate additional revenue, and you should avoid this type of instructor when you fly.

When you are in the air, you should be doing the majority of the flying, with the instructor demonstrating a maneuver or technique then letting you practice. If you have an instructor that does as much flying as you do during a lesson, you may want to look elsewhere. During a lesson the instructor should critique your progress in clear, concise terms that help you become more proficient at each maneuver. If you do not understand what you are supposed to do, or how something needs to be changed to get a maneuver right, ask. Your instructor should be able to answer your question immediately or get the answer for you once you are on the ground. The flying techniques you learn from your flight instructors will stick with you for your entire flying career. Work hard to learn things right the first time because it is much more difficult to unlearn, then relearn a maneuver.

You should also be certain that the planes you fly are well maintained and in safe flying condition. Flying equipment that is inoperative, or that rarely works correctly, reduces your opportunity to learn to fly in the correct manner. If you spend more time trying to work around radio or instrument failures than you do learning, it may be a good idea to look for another flight school. The flight school should also have a curriculum that they can go over with you before you begin flight training. The curriculum does not necessarily need to be in detailed, written format, but should be clearly laid out so that you know what you will be learning and in what sequence. Even if you are just taking a refresher flight on a particular maneuver, the instructor should go over expectations before the flight so that you know what will be taking place during the lesson.

You should feel comfortable with the flight school and instructors that you fly with. If they do not provide professional service, you will not be learning all that you should or be satisfied with the instruction that you receive. Don't feel limited to just the local flight school. If you are serious about learning to fly, it is worth checking out the schools within a reasonable drive of your home. Finally, ask questions as you learn to fly. Flight instruction is a two-way communication process, and your instructor may not be able to anticipate all of your needs. If you enter flight instruction with a desire to learn and the time to be able to fly on a regular basis, you will find that the time spent is one of the most enjoyable of your life.

AIRCRAFT SAFETY

We have touched on safety throughout the book, but I wanted to take one last time to discuss its importance. No pilot ever begins a flight assuming it will end with problems. We all believe we will arrive safely at our destination airport. But occasionally situations take place that are either out of our control or we set ourselves up for through lack of proper action or planning. When you fly, try to anticipate contingencies, plan ahead, and be prepared. We discussed the poor judgment chain and how through compounding decisions pilots create situations that they can no longer safely get out of. Do not let time pressures, stress, or other factors override your decision-making processes.

Safety comes about as a result of consciously thinking about the actions you take and the potential results of those actions. On several occasions during my flying career, I have made the decision to stay on the ground when the weather appeared to be a potential problem. In each of these instances the people that I was flying were somewhat put out but respected the fact that I put their safety before a poor judgment.

Take the time to get proper weather briefings before you fly, and recheck the weather en route as you fly. Make sure you know that the runways are sufficiently long at your airport of origin and your destination. The runway that worked very well during a cold day in January may suddenly be too short on a hot, humid August afternoon. Knowing the density altitude and how much additional runway you will need is extremely important. The same is true of properly loading the plane you fly so that it remains within weight and balance ranges. Having a forward or aft center of gravity can adversely affect the flight characteristics of the plane.

Before you fly, do a thorough preflight. Make sure the plane is in good working order. This should be followed by a runup to verify that the engine and instruments are also functioning properly. As you fly and approach the destination airport, be sure to use the radio to your advantage, and keep eyes moving as you look for other traffic. Each of these items is common to every flight, and it is not possible to list every safety aspect related to flight. By keeping safety as a priority when you fly, and planning ahead, you can increase the safety and enjoyment of each flight.

As a pilot you are given a tremendous amount of responsibility that you should not take for granted. Your passengers depend on your judgment and skills to safely get them to their destination. Other pilots depend on you to smoothly fit into the airspace that you use and follow standard procedures stipulated by the FAA. Don't take this responsibility for granted. Get additional training, learn the airplane that you fly, and stay on top of important changes in regulations. By continuing to improve your skills and taking the task of being a pilot with the proper mindset, you will be an asset to the flying community. As pilots we want to share the enjoyment we get from flying with others. The children you take flying may decide to learn themselves one day.

For me, flying has been a lifelong passion, and the desire to continue to learn how to be a better pilot has been the fuel for wanting to teach what I learn to others. If you approach flying with an open mind and are willing to commit the time it takes to be a proficient, safe pilot, you will spend a lifetime of enjoyment with the activity. I know of no better feeling than flying among the white, sunlit clouds on a summer afternoon. The world takes on a fresh perspective, and the seemingly increasing pace of life seems to slow for a while. I never tire of the magic of flight; it has been with me since my childhood and never waned as I became an adult. I sincerely hope that you receive the same sense of satisfaction and fulfillment during your flying career.

Bibliography

Beggs, Gene. 1994. "Spinning with Gene Beggs." *Sport Aerobatics*, April, pp. 31–36.

Federal Aviation Administration. 1972. *Airframe & Power Plant Mechanics Handbook*. Washington, D.C.

———. Revised 1994. *Airman's Information Manual*. Washington, D.C.

———. Revised 1980. *Flight Training Handbook*. Washington, D.C.

———. "Introduction to Pilot Judgment." Accident Prevention Program. Washington, D.C.

———. "On Landing: Part I." Accident Prevention Program. Washington, D.C.

———. Revised 1980. *Pilot's Handbook of Aeronautical Knowledge*. Washington, D.C.

Love, Michael C. 1997. *Spin Management & Recovery*. New York: McGraw-Hill.

———. 1995. *Better Takeoffs and Landings*. New York: McGraw-Hill.

Page, Vince. 1994. "Letters." *Sport Aerobatics*, August, p. 31.

Index

Note: **Boldface** numbers refer to illustrations.

INDEX

INDEX

INDEX

lazy-8, 189, 193–196, **194**
leading edge slots, 15–16, **16**
leading edge, airfoils/wings, 10
left-turning tendencies of aircraft, 51–55
    See also yaw
lift, 2, 62
    airfoil and lift, 4, **5,** 6–7
    angle of attack, 4, **5,** 6–7, **8**
    banks/turns, 41
    Bernoulli's principle, 4
    climbs, 39
    flaps, 13–15, **14**
    ground effect, 157–158, **157**
    impact lift, 7, **8**
    induced drag, 3
    leading edge slots, 15–16, **16**
    lift enhancement devices, 13–17
    lift vectors, 22–25, **23**
    Newtonian lift, 7, **9**
    slats, 16–17, **16, 17**
    spins, 89
    stalls, 6–7, 62
lift enhancement devices (*see* airfoils/wings;
    flaps; leading edge slots; slats)
lift vectors, 22–25, **23**
load factors
    slow flight, 57–59, **58**
    stalls, 70–73
    steep power turns, 199–202, **199, 200**
longitudinal axis of airplane movement, 18, **19**
loops, half loop, 10, **11**
lost procedures, 204–205
low final approach, **152**

manuevering speed (Va), 40–41, 47, 70–72, 136
maximum braking, abort takeoff procedure, 184
maximum flap extension speed (Vf), 59
mean camber, airfoils/wings, 10
Muller, Eric, emergency spin recovery, 111–112
Muller-Beggs spin recovery technique,
    111–112, 114

navigation, 204–205, 235–240
    global positioning system (GPS), 239–240
    VHF omnidirectional range (VOR), 236–238,
        236, 238
negative camber, airfoils/wings, 10
Newtonian lift, 7, **9**
night landings, 158–160, **161**
normal final approach, **151**
normal mode spins, 101, **102**
normal stalls, 62
    power-off, 64–67, **66**
    power-on, 67–70, **69**

oil-pressure drop, 210–211

panel layout, typical, **226**
parasite drag, 3
P–factor
    slow flight, 51, 54–55, **54**
    takeoff, 162
pitch angle, 19, **19**
    climbs, 46
    slow flight, 56
    spins, 93
pitot tube and airspeed indicator, 34–35
poor judgment (PJ) chain in accidents, 220–221
positive camber, airfoils/wings, 10
power settings
    approach and landing, 147–152, 155
    climbs, 39–40
    crosswind landing, 180
    crosswind takeoff, 179
    descents, 40–41, 47
    short-field landing, 168–170, **169**
    slow flight, 55–57, 59
    spins, 97, 101, 112
    straight-and-level flight, 37
    takeoff, 162
power-off approach, 152
preflight checklist, 160, 162
propellers
    gyroscopic effect, 53, **53**
    P–factor, 54–55, **54,** 162
    pusher propellers, 3
    slipstream, 52–53, **53**
    thrust, 3
    torque, slow flight, 51–52, **52**
    tractor propellers, 3
pump-driven vacuum system for gyroscopic
    instruments, 228–229, **229**
pusher propellers, 3

rectangular course, 118, 123–126, **124**
relative wind, 4
right-turn departure, 145–146
    See also takeoff; traffic patterns
rollout, short-field landing, 168–169
rolls, 19, **19**
    aileron action, 20
    airplane movement about longitudinal axis,
        19, **19**
    half roll, 10, **11**
    snap rolls, 87
    spins, 86–87
    wake turbulence, 8–10, **9**
rotation effect, crosswind takeoff, 178
rough engine, 210

INDEX

About the Author

Michael C. Love (New Glarus, WI) is a flight instructor, aerobatic pilot, and holds a commercial pilot's rating. He is the author of *Better Takeoffs and Landings* and *Spin Management and Recovery* in McGraw-Hill's Practical Flying Series.